Virtual Design of an Audio Lifelogging System

Tools for IoT Systems

Synthesis Lectures on Algorithms and Software in Engineering

Virtual Design of an Audio Lifelogging System: Tools for IoT Systems

Brian Mears and Mohit Shah

ISBN: 978-3-031-00397-4   paperback
ISBN: 978-3-031-01525-0   ebook

DOI 10.1007/978-3-031-01525-0

A Publication in the Springer series
SYNTHESIS LECTURES ON ALGORITHMS AND SOFTWARE IN ENGINEERING

Lecture #16
Series Editor: Andreas Spanias, *Arizona State University*
Series ISSN
Print 1938-1727   Electronic 1938-1735

Virtual Design of an Audio Lifelogging System

Tools for IoT Systems

Brian Mears

Mohit Shah
Genesis Artificial Intelligence

SYNTHESIS LECTURES ON ALGORITHMS AND SOFTWARE IN ENGINEERING #16

ABSTRACT

The availability of inexpensive, custom, highly integrated circuits is enabling some very powerful systems that bring together sensors, smart phones, wearables, cloud computing, and other technologies. To design these types of complex systems we are advocating a top-down simulation methodology to identify problems early. This approach enables software development to start prior to expensive chip and hardware development. We call the overall approach virtual design. This book explains why simulation has become important for chip design and provides an introduction to some of the simulation methods used. The audio lifelogging research project demonstrates the virtual design process in practice.

The goals of this book are to:

- explain how silicon design has become more closely involved with system design;

- show how virtual design enables top down design;

- explain the utility of simulation at different abstraction levels;

- show how open source simulation software was used in audio lifelogging.

The target audience for this book are faculty, engineers, and students who are interested in developing digital devices for Internet of Things (IoT) types of products.

KEYWORDS

virtual design, virtual platforms, integrated circuits, Internet of Things, IoT, system on Chip, SoC

Contents

Acknowledgments

This project was supported in part by Intel Corporation. Thanks to Professor Andreas Spanias at the Arizona State University SenSIP center who provided valuable advice and guidance in all the work involved from inception through to this book. Thanks also to Professor Chaitali Chakrabarti for providing valuable technical advice during the course of the audio development work.

Brian Mears and Mohit Shah
May 2016

CHAPTER 1

Introduction

1.1 VIRTUAL DESIGN DEFINED

The latest innovations in inexpensive highly integrated chips, sensors, smart phones, wearables, tablets, and cloud computing are enabling exciting new product opportunities.

For example, in the near future objects in the house will be connected and searchable. A smartwatch that has voice recognition capabilities can enable searching the Internet for videos that can then be displayed on a smartTV using a smartphone as the controller. These types of capabilities are commonly considered to be part of the application domain labeled the Internet of Things (IoT). Using wireless communication and cloud computing devices such as smart appliances or wearables can access huge computing resources and thereby take on powerful intelligent features [1–3].

The key question addressed here is, "What is the best way to design such complex systems?" Many problems need to be solved, from circuit issues to keeping data private and secure. Implementing a distributed system with many computing elements and complex software is likely to encounter many problems. Designers need to identify these problems at an early stage, before much expense has been incurred in creating custom chips and extensive software. To help solve these problems we introduce and explain new methods that we simply call *virtual design*.

Virtual design combines simulation techniques with a top-down design methodology. This was not possible before; digital hardware had to be built first and then programmed. Using simulation it is possible to adopt a top-down design methodology, where high-level system issues get analyzed first and details are added later. Overall, the process reduces the risks involved with designing complex systems.

Figure 1.1 shows in one compact diagram the overall approach we are proposing.

The top-down design methodology is a logical progression from an idea to an implementation (the rows in the figure). We first create a high-level simulation to explore the design space at the concept stage to find system issues and better solutions. Then we incrementally enhance the concept simulation to develop a virtual platform for more detailed hardware and software development. When sufficient confidence in the simulation results are obtained the team can get started on fabricating chips and building hardware. Virtual platform modeling is a simulation technique to develop software for a new product before the hardware exists, typically for a microprocessor-based design (the blue box in the figure).

A typical IoT system consists of several separate components, such as a smart device, smartphone and Internet service. Designing a real product involves developing hardware and/or soft-

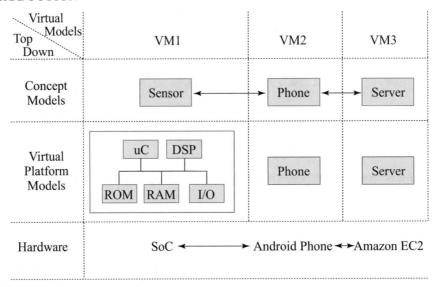

Figure 1.1: Virtual design in one diagram. Rows indicate top-down design from concept models to hardware. Concept models are enhanced into virtual platform models where needed. Columns indicate different parts of the system run in separate virtual machines.

ware for all the components in the system. In virtual design these components are simulated using virtual machine technology (the columns in Figure 1.1). Virtual machines isolate the simulation models of the separate components and enable convenient operation on a single desktop computer (the green boxes in the figure). For example, for the specific case discussed in this book, these components are a wearable sensor, smartphone, and an Internet server. Each of the three components has a simulation model running in a separate virtual machine.

Virtual design allows product ideas to be explored more thoroughly and conveniently prior to building real hardware. To show how these techniques work in practice, we applied them to new ideas for an audio-based product concept. This book describes the design of an audio lifelogging application that involves advanced machine learning techniques to recognize environmental sounds. We describe how virtual design techniques were used to explore the system-level aspects of the application, partition the sensor hardware and firmware, develop the controlling Android application, and, finally, port the software onto a smartphone and remote cloud-based server.

1.2   WHY VIRTUAL DESIGN

Design complexity is the problem. Chip circuit density is now at the point where a complete digital system can fit on one chip and this has led to the term System on a Chip (SoC). SoCs are relatively inexpensive and ubiquitous. A large SoC can contain several processors, digital signal

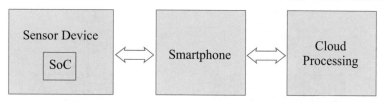

Figure 1.2: Typical IoT configuration.

processing (DSP) blocks, memory blocks, radios, and various interfaces to connect to the outside world. Wired and wireless communication link these devices with each other and to computers on the Internet.

For example, a sensor [4] (audio, video, pressure, temperature, etc.) connects with a smartphone, which in turn communicates with computers in the cloud (Figure 1.2). Behind the scenes can be a large amount of software: the local sensor processing, application code on the phone, and software services in the cloud. The challenge is how to design the components and software without being overwhelmed by the complexity. The rewards can be significant. However, the risks are also great, such as having to redesign a chip in the sensor subsystem, suffering a significant delay in getting a product to market, or loss of sensitive customer data.

Consider attaching a light switch and a house porch light to the Internet. The homeowner can perform a variety of different actions, such as the light can be turned on remotely by the homeowner from his smartphone. The light can be turned off automatically at dawn. The power consumed can be monitored. The basic hardware and software in the SoC is relatively straightforward to implement these days. What is more challenging in this case are factors such as security, privacy, and reliability. Is someone monitoring your switches to see if you are at home or not? What happens if the software has a bug, am I in the dark? Encryption software and other safety measures would be needed to prevent others from monitoring or controlling the lamp. So what started as a simple idea turned into a complex system problem. Consequently, better tools and design approaches are needed to help fix system issues at the start of a project rather than later.

Traditionally, one would design the hardware, write the software, plug it in, and see if it works. It is a bottom-up approach. Typically, the software could not be developed before the hardware was ready, as there was nothing to test it on. Also, the hardware tended to use general-purpose microprocessors or microcontrollers and so fixing hardware problems could often be accomplished by changing EPROM or FLASH memory code. However, the new custom SoC is a more complex beast in comparison. SoC is by definition everything on one chip and cannot easily be changed once fabricated. SoCs often store code in on-chip ROM, which is very expensive to change. Hence, it is desirable to get that on-chip code fully debugged before going into production.

Field Programmable Gate Arrays, better known as FPGAs, have become popular as a way to debug new chips prior to production, however, they have their problems. Most important in

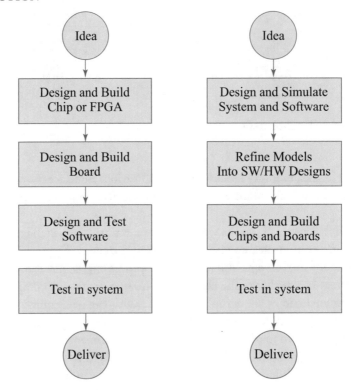

Figure 1.3: High-level view of bottom-up and top-down design methodologies.

this case, it is still a bottom-up approach to design. Hardware design has to come before FPGA testing, then software, then system integration. Generally, a design has to be almost complete before it makes sense to set up one or more FPGA boards.

Hence, to improve on this situation we advocate a top-down design approach using virtual design techniques. Recent advances in virtual machine and virtual platform software make virtual design a more viable design methodology. In this approach we start with a high-level software model to find the high-level issues first, and then gradually add more detail to flush out lower level issues. Figure 1.3 contrasts conventional bottom-up approaches with the top-down approach.

Because one is starting at a high level with initially simple software models, the focus can be on system behavior and not on the details to achieve the functionality of that behavior. For example, using the light switch concept above one can ignore low-level message-passing details to start with and just focus on the more important security concerns. Then gradually add more detail and work toward a more complete implementation, all the while still working in the virtual world. When sufficient confidence has been obtained in the project viability then hardware development can start. The model and the application software developed in the virtual world become part of the specification to design the hardware.

1.3   SIMULATION TECHNOLOGIES

Virtual design is a simulation-based methodology. Simulation of hardware design for software development has been used for many years, largely hidden inside large companies where they developed their own proprietary tools and techniques. What is different now is the availability of standard tools, languages, modeling standards, and methodologies that work through the different design stages to simplify SoC design. Several of these tools are in open source form and allow one to get started without great expense.

In our approach virtual design is supported by two underlying technologies, virtual machines (VM) [5, 6] and virtual platforms (VP). Virtual machine technology allows more than one virtual computer to share the resources of one physical host computer. For example, one can run a Windows 7 virtual machine and a Windows 8 virtual machine on a Linux host computer at the same time. Virtual machines are useful in the design context to conveniently simulate the various devices in the system, such as smart sensors, wearables, phones, and servers. Without VM technology, each device would need to be simulated on a separate computer, which would make running simulations more cumbersome. Sharing one computer will slow down the simulations but fortunately most modern computers have multiple cores and so each VM can run on a separate core, minimizing the slowdown.

Inside a virtual machine we can use the second technology—a virtual platform model to simulate a complex SoC. A virtual platform simulates the hardware of a processor design and enables software for that machine to be developed on the virtual platform. The instruction set in the new design does not have to be the same as the host computer running the simulation. For example, one can run ARM software on a virtual platform model of an ARM-based SoC that is hosted on a regular PC. Put the two technologies together, virtual machines and virtual platforms, and we can create models of multiple devices and computers. We can then write realistic software that will run on those models and find system-level and software bugs before the hardware is built.

Virtual platform, more specifically, is a general term to describe a software model of a hardware device consisting of one or more processors, peripherals, and memories. It might be a single SoC chip (i.e., a smartphone chip), a board of components (i.e., a prototype sensor development board), or a larger system with multiple processing subsystems. The software model is written to model the functional behavior of the component or board. How much detail is included depends on the abstraction level used (discussed at length later). The model allows software that runs on the hardware to be developed on the software model before hardware exists.

There are several companies producing virtual platform tools, and many others are getting started. Cadence [7], Carbon Design [8], Intel [9], Wind River [10], Mentor Graphics [11], and Synopsys [12] have significant software tools and services that cover this market. SystemC has become a standard for high-level hardware simulation and most vendors provide tools that support SystemC. A simulator for SystemC is available in open-source form from the Accellera website [13].

1.4   SUMMARY

The main advantages of virtual design are:

1. A high-level virtual system model allows prototype application software to be started at the concept stage to investigate the project possibilities and viability.

2. The top-down approach in a simulation environment enables an incremental approach to developing software and hardware before hardware is built.

3. System risks get identified sooner and before hardware is built.

4. The initial application software developed in the simulations become implementation requirements for the hardware.

5. The application software can continue to be developed on the system model in parallel with hardware construction, leading to potentially quicker product introduction.

In this book we discuss an audio lifelogging application that is more complex than web-connected lights and switches. Similar to the lights example, the audio application uses the power of the Internet for computations and distribution. This is used to demonstrate some of the virtual design techniques.

Chapter 2 describes the theory behind virtual design, the reasons it is needed, and the methodology.

Chapter 3 shows how we developed the lifelogging application using virtual design techniques. This lifelogging application captures and stores audio data and uses pattern recognition techniques to search for familiar sounds.

We have used this audio project as an example. However, it should be clear that there are many other product ideas one can foresee that will have similar problems and solutions. Any sensor, appliance, gadget, tool, toy, machine, vehicle, or whatever, that has a connection to the Internet to enhance its capabilities, is a candidate for this approach. For even more complex systems with multiple Internet-connected devices that cooperatively work together, the virtual design approach is probably the best way to handle such massive complexity.

CHAPTER 2

Virtual Design

In this chapter we delve into virtual platform and virtual machine technology that make up the concept of virtual design.

2.1 INTRODUCTION TO VIRTUAL PLATFORMS

Simulation of computer hardware and software has been around for a long time. It is only in more recent times that platform simulation has become a significant part of the computer design process. The increasing complexity of computer hardware has made simulation more necessary, and fortunately the increased performance enabled by modern computers has made simulation faster and more convenient. Techniques have been developed in recent years that avoid simulating unnecessary detail and allow much to be done on a fast desk-top computer. It is these areas we focus on in the book.

Virtual platform technology was developed in the semiconductor industry to make support software available as soon as possible after a new chip is introduced. A new chip requires a range of software tools, such as operating systems, real-time kernels, boot loaders, debuggers, test programs, and other code that has to run on the new device. Previously that software development often had to wait until the new chip was available. Providing those software tools sooner reduces the time it takes to get customer products into production.

The term virtual platform applies to a piece of software that models the behavior of a digital system consisting of processors, memories, peripherals, and interconnects. Think of the word platform in this case as the motherboard inside a computer, that has all the essential parts to make the computer work. The platform could also be a mobile phone board, tablet board, or an embedded board of electronics in some piece of electronic equipment. The simulation can also include peripherals external to the board, such as displays, disk drives, or radios. See Figure 2.1.

Most important is the virtual platform simulates the software that runs on the hardware platform. The purpose of a virtual platform is to find software and hardware problems prior to the manufacture of the hardware product. It may not find all problems and often needs to be used in conjunction with other methods. However, it provides a convenient way of finding a large number of issues that would be difficult to find by other means. It is especially important now that highly integrated SoCs are used to build platforms, since the cost of re-manufacturing an SoC to fix a problem can be very expensive.

The basic development flow is shown in Figure 2.2. The first step is to prepare a set of component models, for example, to represent the diagram in Figure 2.1. Models with modular

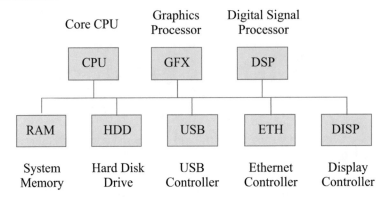

Figure 2.1: A virtual platform diagram.

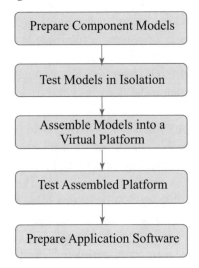

Figure 2.2: Virtual platform development flow.

interfaces can be reused in several virtual platforms, just as hardware components can be reused in different circuits. The models are tested in isolation then assembled into a virtual platform and debugged with platform test programs. Once constructed, the virtual platform can be used for application software development.

There has been a gradual convergence to standard tools and methodologies for hardware simulation. Prior to this period there were many different simulators and different modeling languages, which meant a model developed for one simulator could not be combined with models developed for other simulators. Either the model had to be rewritten or a software wrapper was required to convert from one modeling interface to the other. This is a problem for organizations that have partners and customers that want to use the virtual platform models for their own early code development purposes but use different tools.

The advent of SystemC has been the catalyst for simulation convergence. SystemC is a C++ class library that provides simulation functions for hardware modeling in a C++ programming environment. Being an open-source language, SystemC makes it easy for developers to get started without the expense of a commercial software package. Over time, simulation tool vendors adopted SystemC and have added their own differentiating features, such as graphical interfaces for model creation and debugging. Consequently, a user can start off with the open-source SystemC tools and then move up to more powerful commercial tools. System Verilog is potentially another alternative for higher level hardware simulation, however, it is not targeted for this purpose.[1]

2.2   COMPLEXITY

Here we expand on the issues of complexity and why design has become so difficult. This has been the driving force leading to virtual platform techniques.

Complexity is broken out into two aspects, device related and system related. In the general sense, device complexity involves the hardware and software that goes into the box. The system complexity is what the device creates or has to deal with.

DEVICE DESIGN COMPLEXITY

We take it for granted that our electronic gadgets work flawlessly and it is a big achievement that they largely live up to that expectation. A relatively old device now, the Intel® Pentium® 4 microprocessor was developed in the late 1990's and a report [14] on its validation process shows how difficult it was to design and debug. Note the Pentium is not a SoC, it is just a microprocessor, albeit a very sophisticated one. It took a large team of validation engineers several years to fully debug the logic design. The chip was implemented in 0.18um silicon technology, but since then silicon geometry has shrunk to 22nm and below. That is about 64 times as dense! Consequently SoCs designed now can be considerably more complex.

SOC DESIGN ISSUES

There are several compounding factors that increase SoC design complexity [15] and are significant for IoT applications: power consumption, data security, cost, RF circuits, and analog circuits.

---

[1]System Verilog adds C- and C++-like features to Verilog to primarily support silicon verification. System Verilog is a more complex language and at this time there is no open-source version. Since SystemC is just a C++ class library it is straightforward to compile with other C++ programs. System Verilog can import C routines through its Direct Programming Interface, however, that requires structuring the program as a System Verilog program that calls C functions. At this time it is primarily SystemC for higher level simulation projects and System Verilog for detailed hardware design and verification.

Power consumption is critical for battery-powered devices, and in large systems lower power consumption reduces electric bills and cooling costs. Typical power management techniques [16, 17] basically slow down non-critical functions and turn off circuits not currently active. Another common technique is to design more power-efficient custom circuits for specific tasks, such as for encryption algorithms, which can require exponentiation and modulo calculations on very large numbers. Verifying a power management scheme works as predicted can be quite complex for large chips.

Data security is important for many applications, such as financial transactions. Software encryption provides a basic level of protection, however, high security applications (i.e., military) could require more protection. For secure generation of cryptographic keys, a trusted platform module (TPM) [18] is included on computer motherboards. TPM is a device that contains a dedicated embedded processor that is used to ensure the integrity of the platform. The device consists of key generators, unique root keys, control processor, and firmware. A TPM can also be integrated into a SoC. Another vulnerability is the data that transfers between the memory subsystems and the processors on a printed circuit board, which could be monitored by a malicious user if they gain access to the hardware. In this case, on-chip encryption techniques can be used to protect highly sensitive data [19].

For IoT to spread into our everyday appliances, the cost of an SoC is an important factor. Various techniques can be used to reduce cost, for example, large space-consuming memories can sometimes be reduced by doing more calculations on-chip.

Many of the functions used to implement radio communication can now be implemented in silicon [20, 21], resulting in fewer separate discreet RF components. Combining those radio functions on the SoC can save cost, but at the expense of increased complexity.

Analog-to-digital and digital-to-analog converters are desirable on-chip for sensors, audio, video, and RF to reduce cost, however, getting high-precision converters to work in the presence of large amounts of digital noise can be a big challenge, and that requires specialist mixed-signal designers and techniques [22] to handle the task.

SYSTEM DESIGN COMPLEXITY

In the past, the component designer has left the system issues to others to resolve. The component designer expects to get a specifications sheet for the component and then sets about designing the part. The problem is the system issues are hard to identify in advance. For example:

- Software complexity: Devices often connect to the Internet using a wired or wireless adapter. This can give a relatively simple device access to vast computing resources and thereby create limitless possibilities for bugs to show up. A software feature yet to be implemented and running in the cloud may imply a change to the hardware. For example, the Internet may not respond as fast as expected, requiring more buffering on-chip or a different processing algorithm.

- Security: The hardware described in the last section is just part of the problem. A rigorous methodology has to be employed to ensure software written for the device is indeed secure, and that may even entail security clearance of the development team!

- Safety: An IoT device might be responsible for monitoring a hospital patient and delivering a medical drug intervention. Understanding safety requirements in this situation is imperative.

- Ergonomics: Users have to come to expect good design and want their gadgets to be easy to use and intuitive. In the early stages of product design these aspects are often hard to foresee.

- Radio communication: Standard radios such as Bluetooth or WiFi often operate in the unlicensed 2.4GHz band. Microwave ovens also work in the same band and leaking radiation can stop data transmission completely. This could be a serious problem for a medical device. In such situations a custom radio working in a different frequency band may be necessary. Custom radios are also attractive for long-range and very low-power applications, however, designing a custom radio is far from easy.

2.3   METHODOLOGY

In this section we explain the motivation for a top-down methodology compared to a bottom-up approach. We discuss how the top-down approach starts with simulation models at a the concept level and leads into to more detailed functional models.

ISSUES WITH BOTTOM-UP DESIGN

The current method of designing new hardware and software systems can be classed as bottom-up, i.e., design and build the hardware, add the software, test in the system. Until the advent of good simulation technologies there was not really any choice. It is too hard to write the software without having something to test it on. A prototype hardware board would be built, software loaded and then debugged. This is an acceptable method when the system is not that complex and the new hardware is built from standard components, such as logic chips, memories, and microprocessors. However, as we discussed in the last section, developing SoCs is not that easy.

What happens if the chip is designed first? When the prototype board is plugged into the Internet all kinds of unpredicted system issues are going to show up. If the board is highly programmable then it may be possible to overcome the issues with reprogramming, however, if a redesign is required it could mean a complete new chip.

A separate and perhaps more important issue is the lack of flexibility to explore new ideas more fully before committing to hardware. For complex boards and SoCs, someone will write a requirements specification for the new hardware. It is therefore the spec writers that must be aware of all the intended applications, how they will work, and the implications for the hardware.

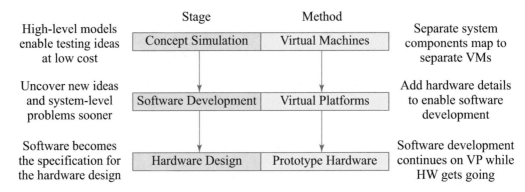

Figure 2.3: Top down design.

This is very hard to do for a brand-new product. This leads to product optimization over a number of iterations, where subsequent versions are required to fix problems and reduce cost. If one could see the requirements more clearly in advance, then product iteration cycles could be saved.

A related aspect of needing hardware and custom chips to prepare software is that the new project is unlikely to be approved by management until there is high confidence of success. By exploring ideas in the virtual space, at lower cost, more speculative ideas can be evaluated and pursued if attractive. Consequently, product design can get started earlier, products introduced before the competition, and revenue realized sooner. That is the potential benefit of top-down virtual design.

TOP-DOWN INCREMENTAL DEVELOPMENT

Figure 2.3 shows the basic top-down methodology. Consider the left side of the picture, which represents the basic stages involved. Start with simple models to represent the concept that is being tested. Debug the concept model and determine if the idea is worth pursuing. The concept model can become a demo tool for selling management on a new idea. If the response is positive, it is likely the demo will inspire new ideas and further thinking. Potential problems are also likely to arise, however, the more problems you can find at this early stage the better, so actively look for them! Finding and fixing system-level issues later in the design process will cause far more grief. Refine the concept model and use it to include new ideas and overcome high-level issues.

The next step is to refine the models so high-level application code can be developed and included. At this point the hardware and software architecture start to take shape. VP models are used to model hardware details that are not present on the host. For example, a microprocessor with a different instruction set or a custom chip. The VP models contain sufficient detail for software development and many simplifications are made so the VP can run as fast as possible. Initially the enhanced models should replicate the basic capabilities of the concept simulation. From there, start adding features to create a more realistic software architecture for the project.

System Level:
    Used for concept exploration

Emulation:
    Used for application development

Functional:
    Used for "bare metal" software
    development, ex OS, drivers

Architecture:
    Used for performance tuning

Cycle Accurate:
    Used for detailed verification

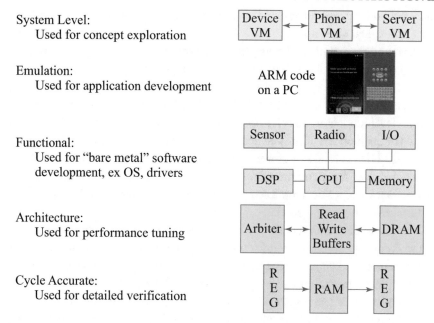

Figure 2.4: VP abstraction levels and representative diagrams.

For example, a real time kernel (RTK) may be needed in the sensor device to manage the software functions. Much of this can be fully debugged in the VP world.

As the software comes together, the green light can be given to start the hardware work. The work done so far becomes part of the specifications for designing the hardware, since the hardware has to implement what the simulations have been assuming.

Hardware design does not have to wait until the software development phase is complete, the two phases can run in parallel for much of the time. As the prototype hardware comes together the software can be used to test the hardware. The audio lifelogging application development involved several stages which are described in Chapter 3.

2.4   VP ABSTRACTION LEVELS

The abstraction level determines the amount of detail in the models of the hardware components. The more detail the more accurate the simulation, but the slower the simulation runs. Simulation speed is important when complicated application code has to run on the simulated hardware, such as when booting an operating system. There are various tricks that can be employed to avoid unnecessary simulations and some of these will be described later. Finding the right balance of accuracy and simulation speed is part of the challenge in building a good virtual platform.

There are now various recognized abstraction levels for hardware modeling: functional, architectural (also-called cycle approximate) and cycle accurate. Here we also add system level and emulation abstractions into this discussion. The levels tend to overlap one another in terms of what they are used for and how they are written. Names tend to vary however we use descriptive or well-known names. In this sub-section we give a brief overview, more details follow in subsequent sections. Figure 2.4 shows the abstraction levels and simple example diagrams.

System-level models are for concept development. They are independent of target architecture, have no timing, and are of limited accuracy. They can range from one program to implement an algorithm, to UML simulators for a large system. Many high-level simulators use block diagram editors to enter the structure of the system and assign simple component model code to each block. In our case, we develop a system model methodology around communicating virtual machines (as shown in the diagram).

Emulation models are typically used for application development, one of the most well-known is the Android Emulator used for testing applications that are written for Android-based smartphones, tablets and smartwatches. Apple also has an application called Simulator, which is used for developing iPhone, iPad and Apple Watch software.[2]

Functional models are a general category distinguished by the lack of timing or having only simple timing. They are commonly used for pre-silicon software development, such as for preparing test programs, BIOS/boot loaders, OS device drivers, and porting an OS to a new processor. Generally, the functional model needs to be true to the data sheet for programming the device, which specifies the instructions, registers, and behavior. Figure 2.4 shows a simple functional model diagram of a SoC.

The architecture model is used for performance optimization and has a lot more timing detail. It is not necessary to have an architecture model for every section of the system, only for those units where performance is critical and cannot easily be calculated. An architecture model contains more detail than a functional model but less than a cycle-accurate model. It runs slower than a functional model but faster than a cycle-accurate model.

Cycle-accurate modeling is the most accurate model and is used for detailed verification. It contains almost as much detail as the Register Transfer Level (RTL) code. RTL code is generally written in a language such as Verilog or VHDL, which can be synthesized to silicon.

Virtual platforms can use different abstraction levels, although most often it is at the functional level. Abstraction levels can be mixed, however it is harder to implement a detailed model with a more abstract model. Low-level detail will be slow to simulate, which can make a large platform model impracticable.

The next few sections will give more insight on how the models are written to implement these abstraction levels.

---

[2]The words emulation and simulation are often used to mean the same thing. Emulation includes models of hardware devices, whereas simulation generally means software simulation only. Emulation is also used to refer to systems made with FPGA's or other hardware that mimic the behavior of a real device.

2.5   SYSTEM SIMULATION

There are various tools for system-level simulation. Perhaps the most well known is UML [23] and various companies provide tools to help draw UML diagrams and assist with code generation. Various free tools are available for drawing the diagrams. However, code generation tools are generally commercial products. UML can help structure a large project and is valuable for organizing a team of software developers. UML may be appropriate to consider prior to setting up a virtual platform environment, and as a way to create some initial code.

Various companies provide block-level diagramming and simulation tools for designing algorithms and system applications. Intel CoFluent Studio [9] is one such commercial product that uses this approach. Block diagrams of procedures or components and their sub-blocks are entered to represent sequences of operations. From the diagrams it is possible to generate SystemC code for fast high-level simulation (SystemC is described in Section 2.11).

In our approach system models are high-level models written in whatever language is convenient. Each component model communicates with other models via IP network message passing and so the language used must have the capability to send and receive IP messages. The models are written and tested separately on a desktop computer. Each component model is then downloaded into a virtual machine (VM) that runs on the same PC. For example, in our IoT application the device model runs in one VM, the phone model in another VM, and server processing in a third VM. See Figure 2.5. Since the VM can provide the same environment as the host, the model does not have to change.

2.6   VIRTUAL MACHINE TECHNOLOGY

Virtual Machine usage has become more common in recent years. It is the basis of cloud computing and has become attractive because native virtualization instructions in the processor core support VM operation. Virtualization instructions enable the performance of a program running on a VM to be almost as same as the performance running on a physical core. Furthermore, modern microprocessors consist of multiple cores, and so it is possible to arrange VMs to run on different cores for best performance.

A virtual machine is a software version of the real machine. Many virtual machines can run on the same physical machine. Each virtual machine can be running completely different software, say Windows 7 on one, Windows 8 on another, Ubuntu Linux on another. From a cloud computing perspective this provides more efficient use of computers. We are taking advantage of VM technology on a PC to create an environment for system simulation, where each subsystem is in an isolated virtual machine and has its own operating system.

The software that creates the VM is known as a hypervisor and is used to create virtual copies of the host computer. See Figure 2.6. Basically, the hypervisor shares the resources and timeshares the host cores so that a user of a virtual machine thinks they have a complete machine. Each virtual machine is independent and can run different software. Thus, if we have multiple

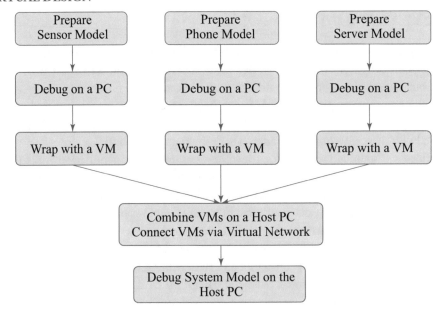

Figure 2.5: Development process using virtual machines.

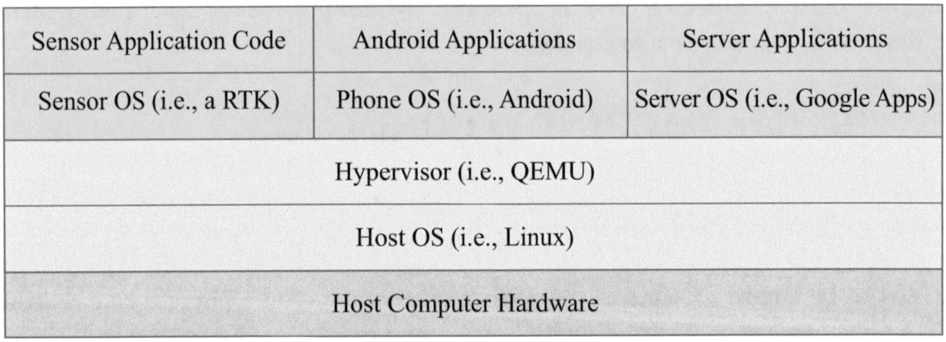

Sensor Application Code	Android Applications	Server Applications
Sensor OS (i.e., a RTK)	Phone OS (i.e., Android)	Server OS (i.e., Google Apps)
Hypervisor (i.e., QEMU)		
Host OS (i.e., Linux)		
Host Computer Hardware		

Figure 2.6: Software stack using VMs.

component models to simulate in our system, virtual machine technology can be used to simulate all of them on one host computer. This avoids having to wire up separate computers to model multiple models.

There are a number of virtual machine suppliers. For our work we used QEMU, primarily because it is a popular free open-source project. QEMU [24, 25] is a type of hypervisor that enables VM creation on a standard PC. Various functional models of different computer architectures have been developed for QEMU and are part of the QEMU package. In our system-level simulation we are time sharing 3 virtual machines on one physical computer.

QEMU can work in conjunction with a program called KVM [26]. It is the two together that provide the high-performance VMs. KVM only works on an Intel based PC running Linux, other programs such as VirtualBox [27] or VMWare [28] are needed for Windows. Intel architecture instructions running in a VM can almost be mapped one for one to the underlying Intel-architecture-based host PC, which is why QEMU/KVM can be so fast.

Communication between VM's is done using TCP/IP sockets. Just as physical computers connect via networks to each other, VM's are assigned IP addresses and can communicate over an internal network isolated from the external world by a bridge interface. The IP network is also used to load the initial program and data for a VM. When using QEMU, a previously generated VM image is used as the base for the new VM. After loading and running a program in a VM the contents of the virtual hard drive will have changed. On termination, the modified contents are stored back in the image file for use next time. Thus, one does not have to start from scratch each time.

For our work we used an earlier version of an open-source program now called QBOX from a company called GreenSocs [29, 30]. QBOX is a combination of QEMU, processor models, and SystemC. QBOX currently supports an ARM9 processor, a selection of ARM peripherals, and the ability to add your own peripheral models in SystemC. They have done the hard work of integrating SystemC code with QEMU.

CONTAINER TECHNOLOGY

In recent years, a new virtualization technique has appeared and is growing in popularity. Linux Container(LXC) [31] technology uses features added to the Linux kernel that enable the creation of light-weight virtual machines. Each container can have its own operating system. However, they all share the same underlying Linux host kernel, and therefore the virtual OS must be a Linux variant, say Ubuntu, Debian, or Fedora. Compared with a traditional VM, container code is smaller because it is sharing the Linux kernel and faster because it avoids the overhead of hardware emulation.

The main driving force for developing containers was not for our purpose of simulation, but for the ease of distribution of complete applications. Traditionally, installation of a new application requires installing other supporting applications and ensuring correct versions of those programs and supporting libraries. By bundling everything required for a new application into a container, it is easier to copy, distribute and install. With less overhead than a pure virtual machine, containers are becoming an attractive option for this purpose.

LXC supports networking so containers can talk to other containers via standard IP messaging. Hence, bigger systems can be built from a set of application containers that communicate via a network interface. Containers can also be nested. For these reason, containers map well onto cloud-computing services. Microsoft has recently announced Windows Server Containers (April 2015), which indicates the growing importance of containers.

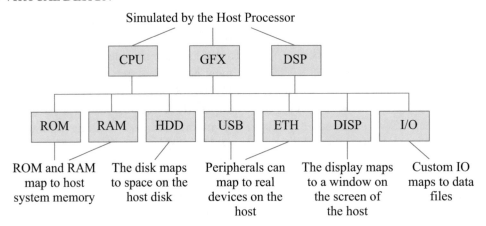

Figure 2.7: Virtual platform diagram of a computer.

From a modeling perspective containers serve two purposes. They are an alternative to VM's for processor subsystems, and within a processor subsystem container there can be further application containers. For example, a model of a cloud service might be a container to represent a server, which in turn contains further nested application containers.

However, containers cannot be used if the code inside needs to access the host hardware resources directly, requiring a bypass of the Linux interfaces. In these situations a traditional VM must be used. VMs and containers can work together in the same machine connected via network interfaces.

At the time of our work, LXC was not as mature as VM technology and was not used for our project. However, for projects that will use Linux, LXC should definitely be considered.

2.7 EMULATION

The diagram in Figure 2.7 shows the structure of a virtual platform used for software development. The boxes in the diagram represent the organization of the target machine to be simulated. The simulated components map onto the physical resources of the host computer that runs the simulation. In many cases the mapping is obvious and straightforward, in other cases it is more challenging. The CPU core model is a simulation of the fetching and execution of the instructions that are used in the target architecture. Likewise, the Graphics controller and DSP blocks generally involve simulation of those specialized instruction sets. To simulate RAM and ROM, part of the system memory can be reserved by declaring a data array in the simulation program. Likewise, a disk drive maps to a file on the host system. More involved is the mapping of target peripherals such as USB, Ethernet controller, and display controller to the host hardware.

The main task of software emulation is to map the target instruction sets of the cores (CPU, GFX, DSP) onto the host instruction set. ARM architectures are common choices for embedded processors in sensor devices and, since the host is mostly a PC of some kind, it uses the Intel instruction set architecture. Therefore, for modeling an ARM-based device a translation must be made from ARM instructions to Intel architecture instructions. When the target and host have similar instruction sets this task becomes much easier.

The simplest translation method is to decode each target instruction into a set of host code instructions that perform the same operation. To avoid the translation process every time target instructions are executed, segments of translated code can be held in a temporary caching memory. Subsequent execution of repeated target instructions is then faster. This process and associated techniques are known as binary translation.

For disk drive emulation it is desirable to pass through the commands to the host, so that the virtual platform can communicate with the host disk drive. In this way, files can be downloaded into the virtual platform from the Internet.

Another emulation task involves mapping the target display onto a window on the host display. This can be done in various ways. If the emulation only has to be compatible at the software library interface, then a thin layer of software can be added that maps target library calls into host library calls. If the emulation has to be compatible at the driver interface, then a more complex layer is required to map to the host driver, since the driver interface exposes more of the underlying hardware features. If the emulation has to match at the register level, then the task becomes one of creating a functional model.

Similarly, other peripherals can be mapped to the host computer. For example a Flash drive. A file on the host disk drive can be assigned to hold the Flash drive memory. The target touch screen or stylus can be mapped to the host mouse. Or for more realism it is possible to access a real USB device on the host, such as a USB mouse, USB Flash drive, USB camera, or USB hard drive from the emulated machine. Special USB drivers map the target USB commands into host USB commands and process the data transfers.

When a target peripheral is completely different from what the host can provide, a simple technique is to map data transfers to a file on the host. For example, a pressure sensor or accelerometer. Data is collected from another device and stored in the file, then accessed by the target when it wants to read the sensor.

2.8   FUNCTIONAL MODELS

Many of the techniques discussed in the last section are applicable here. For functional modeling the main emphasis is on software development of low-level code that controls the hardware directly. This generally means modeling the functions controlled by hardware registers that can be accessed by software. In contrast, emulation models for application software development that only access operating system utilities or system libraries can use shortcuts that avoid register modeling.

A functional model is required to model the correct functional sequence of operations but not necessarily the correct timing. For example, CPU interrupts should come in the right order, however, not necessarily on the same CPU cycle that would be seen in real hardware. In most cases this is sufficient accuracy for software development. User visible register detail must be included in the model so that software device drivers can be simulated. The device driver sets bits in device registers to control data transfer modes and other parameters of the device.

Autonomous units can have many registers. The software driver developer needs to write the code that configures them. Not all registers need to have the functionality associated with each control bit, it depends on the use case. For example, certain test mode registers may not be needed for operating system driver development, since they are only used by test software.

Implementing the model therefore requires implementing the features behind the key control registers. Simple ways are found to represent the functionality. For example, the transfer count register in an ethernet controller is simply incremented by 1 for every byte transferred. There is no need to model the count increment logic, since all that we care about is that it has the same behavior. The actual hardware may involve hundreds of silicon devices to implement that function, however, it is a don't care for the software developer. Many details, such as buses, caches, buffers, memory controllers, synchronization logic, pipeline logic, and others can be ignored when preparing a functional model. Only if the programmer can do something to a block that will change the functional result does the model have to implement that functionality. In many cases only a few control bits in the register have an effect on the simulation and the rest can be ignored.

Data transfers between blocks can also be modeled in simple ways. The bus detail can be skipped. A real bus may have many control signals that synchronize the data transfers. The hardware has to go through an arbitration phase to get access to a common bus, the address and data have be sent in a proper sequence, and responses sent back to signal successful transfer. Fortunately, in a functional model most of this detail can be ignored. It can be assumed the transfer is going to work and therefore the bus transfer can be reduced to a procedure call that just transfers the data. The initiating module calls a procedure in the target that performs the transfer. The return value from the procedure call can be the status of the transfer. In summary then, only details that result in program visible changes need be modeled.

2.9   ARCHITECTURE MODEL

An important area to model in a SoC is the performance of any on-chip caches. Fetching instructions from off-chip memory is relatively very slow, however, on-chip the close connections to the cache enable high-speed execution. This is often an important case to simulate to see how well the proposed architecture performs with different programs. The diagram (Figure 2.4) shows a small section of a simple memory controller and external DRAM memory. The models would be at this level of detail and include timing delays of data passing through each unit.

Figure 2.8 shows the basic difference between a functional model and an architecture model. The architecture model (lower diagram) has a detailed cache model to speed up the CPU operation. The cache behavior depends on the execution flow and what was previously loaded into the cache. The cache policies (for example, write-through or write-back) and memory structures are

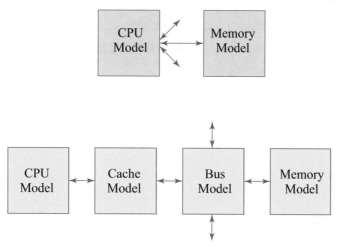

Figure 2.8: Functional model (top) vs. architecture model.

modeled functionally and timing delays are approximated as a set of delay times in numbers of clock cycles. There may also need to be a bus model since, when the bus is busy, memory access delay can be significant. In contrast, a functional model of a processor with a cache would not bother to model the cache or bus at all, since it does not change the target code execution, only its performance.

Architecture models (sometimes called cycle-approximate models), when carefully written, can come very close to cycle-accurate models in accuracy but run faster. For example, consider the implementation of a timer model. A counter is preset with a delay value and after counting down to 0 will set a signal. A cycle-accurate model will activate every cycle to decrement the count and test for 0. However an architecture model will arrange for the simulator to ignore this module until the time the counter is known to reach 0. All the intermediate clock and counter operations were avoided, thereby speeding simulation.

For modeling timing, the first step is to determine which delays are constant and which are dynamically changing. For example, the delay through a static RAM is constant. It takes so many nanoseconds to access the array and output the data. This can be modeled by a fixed delay, which is often multiples of the clock cycle. In clocked systems the address becomes fixed on a clock edge, and the data is loaded into the output register on another clock edge. Another example is a micro-sequencer executing a predefined set of instructions with no variation in timing. In this case, the total delay can be calculated and used in the architectural model. The more delays that are constant, the easier it is to write an architecture model.

Dynamic delays require more programming to model. A common need is to model multiple units sharing a bus. In order to share the bus, each unit has to arbitrate for permission to use the bus, transfer data, and then release the bus for other units to use. In modern processors the bus

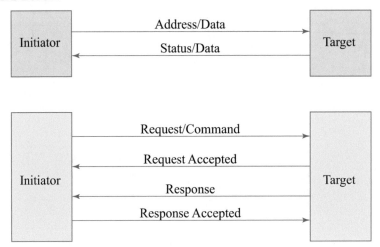

Figure 2.9: Functional handshake (top) vs. architecture handshake.

to system memory is the major bottleneck and many architecture techniques are used to avoid delays of critical data. Similarly, cache operation is very dynamic.

A common arrangement for modeling dynamic situations is to use a message protocol between two units to calculate the variable delays in sending data. Functional and architecture versions are shown in Figure 2.9.

For a functional model there is little or no timing accuracy. In the case shown, the initiating agent (top part of diagram) wants to read or write data in the target device, such as a memory module. It simply calls a procedure in the target module to transfer the address and data values. The call return contains the data read, status indication, and time taken. Status can indicate various conditions such as operation succeeded or out of memory range error.

An architecture model is more involved and a hardware-specific handshake protocol is used. The lower diagram shows a general scheme of coordinating transfers between initiator and target. The forward request and backward request-accepted handshake messages ensure the target is not overloaded with requests. Likewise, the response and response-accepted messages coordinate transfers in the opposite direction.

In practice, model developers work in conjunction with hardware architects to decide which features are needed or which can be ignored. If a feature is rarely used it may not need modeling.

2.10   CYCLE-ACCURATE AND RTL ABSTRACTION LEVELS

Cycle-accurate models are more detailed than architecture models but faster than RTL (Register Transfer Level) models. Cycle-accurate models use a clock to synchronize the sequencing of the functions of a model and are almost as detailed as the RTL. They are used where every clock cycle in the hardware design is important, for example for bus controller design, where one or two cycles of inaccuracy might make a significant difference to system performance.

RTL models are ultimately the most accurate because they are used in synthesis tools to generate the hardware. RTL models when simulated show all the possible operations and behavior. RTL models can be slower because they have to model multi-state signals. In hardware a signal can be left in a floating state, pulled high or pulled low by a resistor connected to power or ground, respectively. Cycle-accurate models generally abstract away that low level of detail. Tools now exist that can extract cycle-accurate models from RTL code, and can offer simulation speed increases of 4x or more.

The diagram (Figure 2.4) shows a simple RAM memory that would have well-defined timing, an address register on the input, and a data register on the output. Under clock control, the data register would be loaded 1 clock after the address has been clocked in. Cycle-accurate models generally model every action on every clock cycle.

The need for cycle-accurate models has been minimized by the availability of fast RTL simulators. Proprietary techniques are used to speed execution, making clock-cycle models unnecessary.

However, cycle-accurate models make sense when tools exist to automatically synthesize those models to RTL. Synthesis tools are a new area of research and development and a rapidly evolving field. They allow the user to write at higher levels of abstraction and use compilation techniques to generate the detailed RTL.

2.11 SYSTEMC

SystemC [13] is a class library used with C++ to create a more convenient software environment for simulating digital hardware. Programs written in C or C++ alone are fine for modeling algorithms. However, hardware can have simultaneous parallel operations and other requirements that are not so easy to model.

SystemC has been developed over several years. Earlier versions were focused on hardware design and validation at the cycle-accurate level. Later, Transaction Level Modeling (TLM, explained more later) was added to provide support for higher-speed functional modeling. SystemC gets used primarily for design verification, virtual platform software development, and architectural exploration. During the industry standardization of SystemC, various companies came together to agree on SystemC features and how models could be structured to be inter-operable. This is important to allow a large model to be more quickly assembled from collections of models from different suppliers. Many companies now provide models, libraries, and tools to support SystemC usage.

A short description of the significant features will be given here to give a flavor of what is involved. See [13, 32–35] for more details on the SystemC language and its use.

SystemC provides many hardware modeling features: different data types to represent different word length registers; separation of a design into modules; message passing between mod-

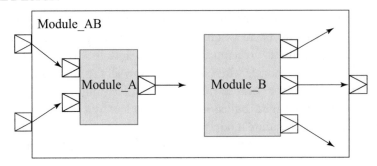

Figure 2.10: Module diagram showing nested modules.

ules; timing delays; synchronization of module execution; debug features; simulation setup; and many other features.

Bit Widths

Digital hardware blocks are connected together by individual signals or busses representing groups of signals. Computer hardware is often thought of as having 32-bit or 64-bit buses and operations, but inside the blocks many different size operations, often from 1 to more than 256 bits, can be found. Hence, to model this detail SystemC provides various data types to represent the different bit widths. Word lengths can be more or less than the host computer word length, which implies long words can require multiples of the native data types.

Modules

A hardware design is broken up into modules, see Figure 2.10. In a high-level design, each module represents a piece of hardware, such as a processor, peripheral, or memory. In a low-level design, a module might be an arbiter, decoder, or arithmetic unit. Modules contain ports that are the interface points with other modules (shown as small boxes in the figure).

Modules are connected through their ports using signals and channels that basically pass data in controlled ways. Modules can be nested, in the diagram Module_AB contains Module_A and Module_B. Internally Module_AB links to Module_A, which is connected to B, and B connects back to AB and other modules inside AB. By nesting in this way, elaborate hierarchies of module structures can be created to represent more complex models.

A module is implemented as a C++ class and class functions are provided to setup and control the module behavior. The module code generally specifies the input and output ports, registers, memory arrays, and functions that will implement the module behavior. Special functions determine how the module is triggered into operation by external events. Other mechanisms, such as sockets, group together the functions needed for inter-module port-to-port communication.

Simulating Parallel Execution

The SystemC scheduler is what coordinates the sequencing of module execution. In real hardware each module is independent and can operate in parallel with other blocks. When simulated in software only one module can execute at a time. The scheduler creates the the illusion of parallelism using event scheduling and a task run queue to determine which module should run next. An event, such as a signal changing value, adds an entry into the run queue to cause a selected module to execute when other higher-priority operations have completed. By arranging the order of events, modules that would run in parallel in the real world run sequentially in the simulation and give the same result.

All digital systems generally use a clock signal for synchronization, and cycle-accurate models are written to simulate each module in the systems for just 1 clock cycle of operation, then repeat again for each module in the next clock cycle. However, a problem with switching from one module to another every clock cycle is very slow simulation speed. It is better to simulate as many clock cycles of activity as possible before switching, as each simulator context switch is expensive in terms of the time needed to save and restore the model state. Architecture and higher level models written in SystemC avoid context switching every clock cycle. They do this in SystemC by using two methods to avoid context switches, event processing and temporal decoupling.

Event processing is for fine-grain control and temporal decoupling for coarse-grain control. Events can be set up to trigger a module to operate once every clock cycle or preferably only when needed. For example the timer unit discussed in Section 2.9. The timer sends a signal at regular intervals to cause some action. Inside the timer is a counter driven by a clock that triggers the signal when the counter reaches a predetermined limit. An exact simulation would increase the count and check against the limit every clock cycle. However, all that is necessary is to set an event to occur at the desired time delay, thereby avoiding the counting task. If there is nothing else to do while waiting, the simulator will skip forward in time to handle the event immediately, thereby saving many simulation cycles. In theory this approach can lead to simulations that are faster than the real hardware, however that rarely happens, due to hardware that is more complex to simulate.

Temporal decoupling is a more aggressive strategy to keep simulating the same module for as long as possible. This is useful for modeling a system with multiple processors or DSPs where communication between processors is relatively infrequent. The first processor module is allowed to run ahead for some predefined time period, known as the time quantum, and then stops itself. Another processor module is then allowed to run to catch up in time. See Figure 2.11. Likewise, all processor modules take it in turns to run ahead and the cycle repeats until simulation ends. In this way there is a maximum time-synchronization error equal to the time quantum value. The time quantum is adjustable and is set between a short value to improve accuracy and a long value to improve simulation speed.

If the time increment is set too long then the two models can get out of sync and cause functional errors. For example, if processors signal one another via interrupt signals it may not

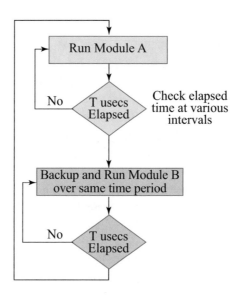

Figure 2.11: Temporal decoupling.

matter too much as to when the interrupt is serviced. However, if the hardware is performing a time critical task then the quantum will have to be set to a low value.

TLM Message Communication and Synchronization

Transaction Level Modeling (TLM) was added to SystemC to improve simulation speed for more abstract models. It is faster than low-level signaling since it skips many of the hardware details. In TLM modeling all we care about is the functional and timing behavior, and that can be simulated in ways that avoid low-level detail. The simplest TLM message-passing mechanism just involves a procedure call from one module to another with data transferred via a pointer. The return from the call can indicate if the transfer succeeded or not and the time taken for the transfer.

In Figure 2.12 three modules are shown connected using TLM message passing. The forward paths between modules represents the procedure call to the next module. (Ignore the reverse paths for the moment.) When the CPU model wishes to read data from the memory, it initiates a call to the router module and waits for the call to return. The router decodes the address in the message and then initiates another call to the selected memory module. The memory module returns the data to the router and in turn returns it back to the CPU. Thus a complex hardware operation has been modeled by a few procedure calls. In this situation the scheduler did not need to switch context and so the operation is quite fast.

For architecture modeling, a more elaborate handshake may be required to accurately model the transfer sequence, like the lower diagram in Figure 2.9 described earlier. If either module is busy it needs to tell the other module to wait. Synchronization requires the target module to

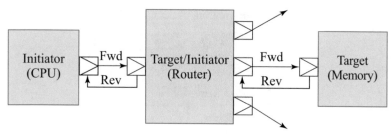

Figure 2.12: SystemC TLM connection diagram.

inform the initiating module it received the request, and the initiating module has to inform the target module when it received returned data. The reverse paths and forward paths shown in Figure 2.12 are used to implement that synchronization protocol.

Other SystemC commands stop and start the simulation, provide debug, and trace features. For example, for architectural design work it is nice to see the system behavior on an oscilloscope-like display, and so, in conjunction with other tools, the trace data can be presented as a time plot of signal and data behavior. SystemC has many functions and features and interested readers should consult the references for further details.

On top of SystemC, design automation suppliers provide elaborate tools to assist in model generation and usage. For example, these companies provide elaborate GUI's to enter the design, debug, and monitor the simulations. Other tools convert and extract high-level models from a low-level abstraction, or synthesize detailed models from more abstract models. New tools and upgrades are being announced frequently. Companies such as Synopsys [12], Mentor Graphics [11], Cadence [7], and Carbon Design Systems [8] are major players in this field.

GreenSocs QBox

GreenSocs provides open-source tools, libraries, and services to support SystemC modeling [29, 36]. We used an earlier version of their current Qbox [30] software to connect applications running under QEMU to access SystemC models.

The main advantage of using Qbox is it provides the SystemC modeling environment with a fast simulation of a microprocessor core and peripherals. QEMU provides the high-speed core simulation and all major processor architectures are supported under QEMU, such as ARM, IA-32, IA-64, MIPS, PowerPC, and SPARC. These are detailed functional models suitable for code development. Developing a processor model in SystemC is a major undertaking and so being able to plug in a QEMU microprocessor model is a big plus. A good use for Qbox is for connecting a processor to a SystemC peripheral to enable development of code prior to hardware availability. The SystemC models are written to include the register interfaces and functional behavior so the users code can set and test the model in the same way as on the real hardware.

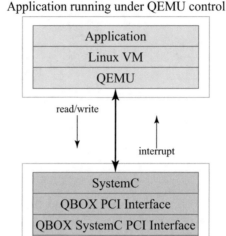

Figure 2.13: Qbox structure.

Qbox has the job of coordinating QEMU and SystemC. Figure 2.13 shows a high-level representation of the structure of Qbox with PCI interfaces. The application code in the core model accesses registers/memory in the SystemC model through a modeled PCI interface. Qbox does the translation from application code to SystemC read/write calls. In the reverse direction, the SystemC model can send an interrupt to the virtual processor core.

2.12 CLOUD COMPUTING AND SERVICES

Developing code using cloud services [37] is an alternative to desktop development. Since the IoT application is going to use the Internet, the server part can be simulated in the cloud since that is most likely where it will run. Generally, cloud-computing services are not free. Thus, it depends on the workload and the budget available as to when to move from the desktop computer into the cloud. Amazon offered free services for a limited time period, and that encouraged us to use those services.

Cloud computing effectively provides many virtual machines by sharing the resources of multiple server computers. Behind the scenes, cloud computing uses similar techniques as QEMU, which shares the resources of one computer, except its on a much larger scale, where large server farms provide enormous compute power to many users.

Amazon EC2 [38] is an example of a cloud-computing service and we used it to host the server software used in the audio application. EC2 (Elastic Cloud Computing) provides virtual machines over the Internet. Each virtual machine can be loaded with a selected Operating System, such as Windows or Linux. Users can then install applications on the virtual machines, which are then accessed over the Internet. The system is "Elastic" in how resources (virtual machines, memory, disk space, compute power, bandwidth) can be dynamically scaled up and down based on the application workload. A range of virtual machine instances are provided, we used a micro instance to prototype the audio analysis code that runs on a server.

From a development viewpoint, cloud computing has the following advantages. Generally, when starting development the data bases and compute requirements are likely to be small. As the application grows and testing expands the required resources grow. The flexibility of cloud processing allows the developer to add these resources and find out directly what is required to run the application. If the application is to be moved back onto dedicated servers, knowing the software requirements can be used to specify the hardware more precisely.

More recently (2016), cloud services are being offered to support IoT development. Microsoft Azure IoT Suite enables block diagram editing of Machine Learning applications [39], and similarly IBM Bluemix [40] supports data and analytics in the cloud.

CHAPTER 3

Audio Lifelogging

To investigate virtual design, a concept topic was chosen that was on the verge of being viable and something we understood enough to tackle. The authors have backgrounds in audio signal processing and so, from discussions with others, a project was devised in the area of audio lifelogging. Many of the speech processing functions had been developed previously for audio work and hence we had a good starting point. The theory will be outlined so readers can get a sense of what is involved and where potential problems might occur. First some background on lifelogging.

3.1 INTRODUCTION TO LIFELOGGING

Blogging or posting about events in one's life has become extremely popular. People are constantly uploading pictures, videos, and their experiences to the web to share with their friends. Lifelogging [41–46], to some extent, can be considered as blogging taken to the limit where all personal data is continuously captured and recorded. The advantage of continuous recording is the user may not have time to get out their phone and take a picture or video, instead events of interest can be extracted from the recordings at a later time. Various people have experimented with head mounted cameras to capture their day-to-day activities. The thought of being able to catch ones complete life and then go back and experience the past is fascinating to many.

Given this general interest, we decided to look at addressing part of the required technology, namely audio capture and searching. A user, equipped with a wearable audio recorder [47], should be able to record audio for a day or longer. During this period, the user might have an interesting conversation with a friend or business colleagues, hear a new piece of music on the radio, or just be listening to the sounds of wildlife while on vacation. At some later point, the user may wish to share these events with friends or simply choose to recall those memories for themselves.

The problem with capturing so much data is going back and searching for events of interest. A text comment could be added manually to allow text searching, but this is tedious in a lifelogging scenario. Instead, it's better if we can have audio searching, i.e., searching for specific sounds in a recording. For example a person has to resolve a legal issue on a accident insurance claim and only remembers the wailing sounds of an ambulance coming to the scene. Or (for a more pleasant memory), a father recalls his children running to an ice cream truck that was playing a distinctive musical tune. Hence, it would be convenient to allow one to search for selected sounds and see what past events show up in their audio lifelog that match those sounds.

When this work was started (in 2008), the notion of lifelogging was somewhat esoteric. It was difficult to predict what users would tolerate wearing to capture the audio. For the investiga-

Figure 3.1: Samsung smartwatch with audio record feature.

tion, a lapel-mounted recorder with local storage was envisaged as the data capture device. Now that smartwatches are available it is more realistic to consider smartwatches serving this function, see Figure 3.1. Current smartwatches have a microphone and a companion smartphone application supports audio note taking. A wireless link connects to the users smartphone to store the audio to a file on the phone. The smartwatch is not currently capable of continuous recording or continuous storage, however, that may be possible in the future.

Lastly, we should say a few words about privacy [48]. The privacy issue was a partial concern in the beginning of the research work but since such devices did not exist back then it was a case of lets try it and see. As we started exploring with the virtual platform it became clearer that privacy would be a concern, and events in the news made it clear this was a sensitive issue. Consequently, we realized the concept is limited to applications that are not sensitive to privacy issues. These are exactly the sort of problems you want to find early when developing new products, uncover the issues in the virtual space and re-target the concept to viable markets.

3.2   OVERVIEW OF ALGORITHMS

A typical block diagram of a lifelogging application is shown in Figure 3.2. The audio is captured with a wearable microphone and compressed to reduce memory space and bandwidth requirements. The compressed audio is transferred to the server where it is analyzed to recognize embedded sounds. Feature extraction identifies key speech features needed for sound recognition. Segmentation breaks up the sequence of features into similar groups, and sound modeling matches the incoming sounds with those in the data base. The new data is stored in the data base for later retrieval. The smartphone is used to control this process and for viewing the results.

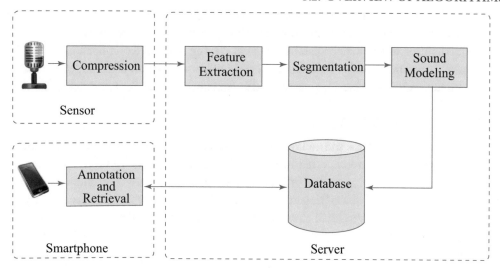

Figure 3.2: Block diagram of a lifelogging application.

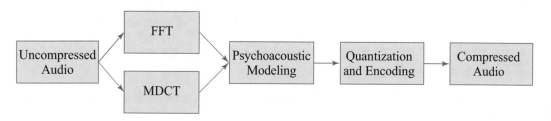

Figure 3.3: Overview of a typical compression algorithm.

A number of different algorithms and implementation techniques exist for each device-specific task/function. In this section, we will provide a brief overview of the different methods. For a detailed explanation, the reader is suggested to refer to our earlier works [49, 50].

3.2.1   COMPRESSION

Audio compression ([51–56]) is used to reduce the storage required in the sensor device, since the device is not continuously connected to the Internet. In this case, the audio was compressed to 64Kbps, a compromise between keeping good audio quality for pattern recognition accuracy, and the memory storage required in the sensor for one day of continuous operation. The audio is uncompressed in the server to enable the pattern matching.

An overview of the most important components of the encoder is shown in Figure 3.3. The raw digital signal first undergoes two kinds of transformations—Discrete Fourier Transform (DFT) and Modified Discrete Cosine Transform (MDCT). These transformations help to shift the analysis from the time domain, where an audio signal is highly irregular, to the frequency domain, which is a more natural way to observe the vibrations manifested in an audio signal. Once transformed, different frequencies can be allocated varying bits based on their audibility. The latter is calculated based on models of the human ear-brain combination, often called psychoacoustic models. Finally, this is followed by encoding (bit packing) to yield the compressed signal.

3.2.2 FEATURE EXTRACTION

Feature extraction transforms the audio signal to a set of features that capture the salient acoustic properties. Ideal features are expected to be: (i) discriminative, and (ii) robust to noise or background sounds. The set of features varies based on the application under consideration. After many studies and for the case of identifying ambient sounds in lifelogging, a set of features that extract various time and frequency-related characteristics were identified. This set includes 6 features—loudness, temporal sparsity, spectral sparsity, spectral centroid, transient index, and harmonicity. Each feature is extracted from short frames (20 ms–40 ms) owing to the non-stationary behavior of audio. For such short periods, many audio parameters can be considered unchanging, i.e., stationary. A detailed explanation related to the extraction procedures can be found in [57] and [58]. Here, we provide a brief description of each of the above features and why they are deemed suitable for lifelogging.

- *Loudness:* The average energy in decibels of a frame. As the name suggests, loud sounds have a higher energy and loudness.

- *Temporal sparsity:* The sparsity of a signal along the time axis. Footsteps in relative silence will have a higher temporal sparsity.

- *Spectral sparsity:* Analogous to its temporal counterpart, describes the sparsity along the frequency axis. Sounds with high background noise or wideband nature will have a lower sparsity compared to sounds like bells or alarms.

- *Spectral centroid:* The center of gravity of the frequency spectrum. Shrill sounds such as breaking glass will have a higher centroid compared to dull, gloomy sounds such as the hum of a machine.

- *Transient index:* Useful in detecting sounds that exhibit constant fluctuations in the spectral content from frame to frame, e.g., crumbling newspaper.

- *Harmonicity:* This feature indicates the extent or degree to which the underlying frequency response exhibits a harmonic structure. Music is expected to have a higher degree of harmonicity compared to crowd sounds or noisy backgrounds.

A few examples of select features extracted from different sounds are shown in Figures 3.4 and 3.5. We can observe how different features respond to each sound and provide a way to

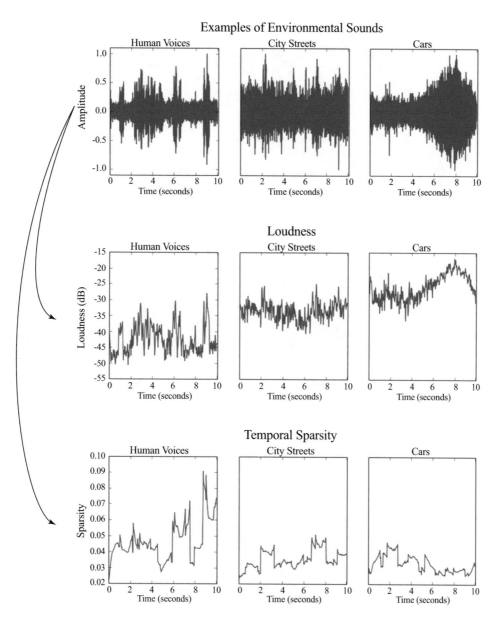

Figure 3.4: Temporal features for different environmental sounds.

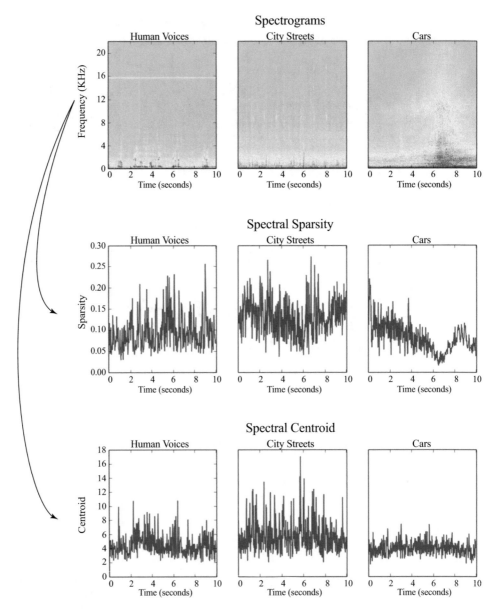

Figure 3.5: Spectral features for environmental sounds.

visually discriminate between different sounds. In certain cases, features for different sounds may be similar. A feature set consisting of diverse features allows one to overcome this behavior.

In Figure 3.4 we see three examples of environmental sounds and the processed results for loudness and sparsity. Overall loudness increases as one would expect going from voices, to city street noise, to a passing car. In contrast, temporal sparsity reduces for city streets and cars. Car noise is much more monotonic. In Figure 3.5 we see the results of the frequency analysis of the three types of sounds. Car noise has a distinctively lower spectral sparsity and spectral centroid. Processing the audio in this manner makes it easier to recognize, although there are no clear distinctions in many cases.

3.2.3   SEGMENTATION

For lifelogging, audio needs to be continuously captured and features extracted over these long recordings (typically, a day). Such recordings would consist of multiple events leading to a sequence of different sounds. The human brain is incredibly efficient at identifying the boundaries between different events based on their sounds, thus allowing us to break down a long recording into smaller chunks and process them individually. We must do likewise in order to build efficient sound models and useful systems. First, we must define a sound event—the continuous segment of audio during which the acoustic properties are homogeneous. Figure 3.6a shows an example recording, which consists of three events occurring over time. One can observe that features of each event (Figure 3.6b) are quite distinct from each other, either in the shape of the feature trajectory or simple statistics like the mean.

The segmentation algorithm used in our work is based on a dynamic Bayesian network (DBN) [59]. DBNs look for changes in the statistical attributes of the features to indicate the presence of an event boundary. An example of such a DBN is shown in Figure 3.6c. This model is capable of identifying the times at which a sound event begins and ends. This information is then used to break down the recording into smaller chunks. See (13) for further details of the algorithm.

The diagram shows three relatively loud sound events separated by quiet periods. The time waveform is processed to extract three audio features (b) and in each we see how the parameter responds to the input. The harmonicity parameter does a good job of separating out the 3rd event, the spectral sparsity identifies well the first event, and the spectral centroid peaks on the 2nd event. Looking at the time waveform it looks as though it should be easy to identify the 3 segments in time, however, simple thresholding systems do not work well in general and so a more complex algorithm is required, in this case DBN. The diagram (c) shows when the 3 events are judged to start and end in the time waveform, which allows them to be separated out (d).

3.2.4   SOUND MODELS

Once the individual events are extracted via segmentation, we attempt to model each event in a manner that allows us to measure the similarity or dissimilarity between two sound events. This

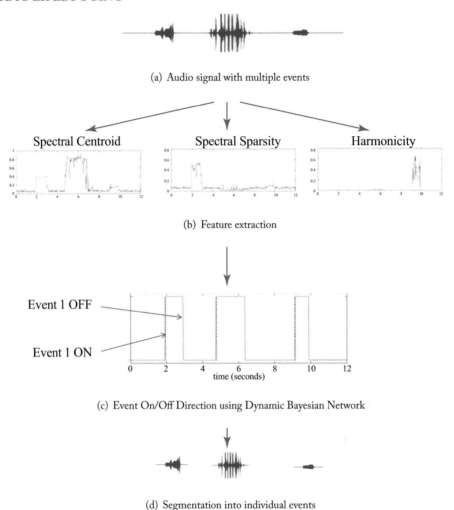

(a) Audio signal with multiple events

(b) Feature extraction

(c) Event On/Off Direction using Dynamic Bayesian Network

(d) Segmentation into individual events

Figure 3.6: Segmentation process in lifelogging.

measure, in turn, would be useful for classifying newly recorded events to one-of-many categories. This, in turn, would facilitate automatic annotation and retrieval, the central aspects of a lifelogging application.

Once again, a number of techniques exist for constructing such models. Here, we present a method that models each sound by comparing it with predefined templates stored in a central database. First, the templates are constructed during training and consist of polynomial curves of varying orders fit to the feature trajectories of the sound events fed during training. Figure 3.7 shows one such example of different order polynomial curves fit to the features. The 0th order

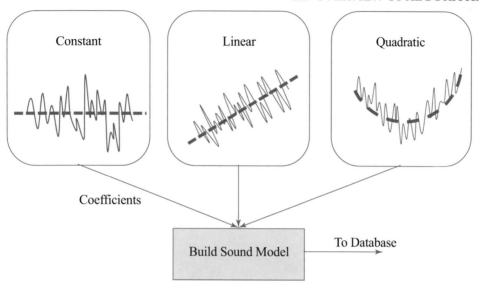

Figure 3.7: Sound modeling process in lifelogging.

polynomial is essentially a constant, which measures the average level of the trajectory and indicates whether the underlying feature is constant or not. The 1st order polynomial characterizes the linear behavior, i.e., whether the feature exhibits an upward or a downward trend. Finally, the 2nd order polynomial models the up-down-up or down-up-down behavior. The results are coefficients of the different curves fit to these trajectories, which are then assimilated to form a template or model for each sound event. Two sound events can now be compared by simply measuring the similarity between their respective templates.

3.2.5   ANNOTATION AND RETRIEVAL

Annotation and retrieval are the key aspects to a lifelogging application. Given the large number of sound events that would get captured daily, it is imperative to design a mechanism that would automatically categorize and annotate these events with minimal human input and reliable accuracy. Secondly, effective mechanisms for retrieval are necessary for a user to be able to navigate through his/her archives.

We constructed a sound-tag network, as shown in Figure 3.8, which stores and characterizes the relationship between sound events and keywords/tags in the form of a graph. The vertices correspond to sounds or tags, while the edges correspond to sound-sound, sound-tag, and tag-tag similarity in the form of weights or costs. The sound-sound weights are obtained by comparing the coefficients obtained via the sound models described above. The lower the weight, the more similar the sounds (two sounds with cars will have a lower weight compared to a car and a person talking). Similarly, the tag-tag weights are obtained using a simple WordNet [60] (tags such as

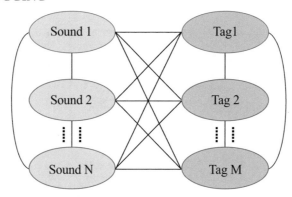

Figure 3.8: Sound-Tag network archived in the database.

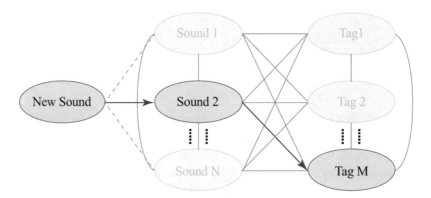

Figure 3.9: Annotation of an unknown sound by traversing through the Sound-Tag network.

talking and voices will have a lower weight as opposed to talking and typing). Finally, the sound-tag weights are learned from the users the only component that involves human input. To restrict such intervention at a minimum, initially a large number of users are asked to assign tags to different sounds chosen from a fairly large set of sound events. This, in turn, is used to initialize the sound-tag weights for each user's respective lifelog.

The process for automatic annotation is shown in Figure 3.9, where a new sound must be assigned relevant tags. These are obtained by computing the shortest path from this new sound to all sounds and consequently the tags in the network. Similarly, for retrieval, the user provides a keyword/tag. In this case, we enter the graph from the right side, as shown in Figure 3.10, and compute the shortest path from the supplied tag to all tags and then all sounds in the network.

The more data that is collected, the more accurate the matching can be. The system has to be tested in many audio environments and with different use cases, i.e., business vs. pleasure. Much of the algorithm tuning work can be done independent of the software/hardware implementation,

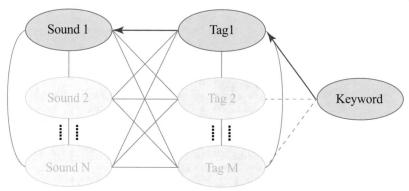

Figure 3.10: Keyword-based retrieval of archived events by traversing through the Sound-Tag network.

however factors such as simultaneous multi-user performance, user convenience, and use-case considerations involve optimizing the algorithms in a systems context.

3.3   IMPLEMENTATION

The system envisioned to implement lifelogging consists of a sensor device, a smartphone, and an Internet service (as per Figure 1.2). The functionality is distributed across the three devices. The sensor facilitates the acquisition of raw audio signals via a wearable microphone. The server performs the computationally intensive machine-learning algorithms for identifying the sounds. The smartphone gives a user the ability to control the sensor, listen to the recorded audio, post it to the web, and search for sound clips on the server. This architecture, although simple, is likely to be useful for other applications, making it a good case study.

The system enables media blogging and lifelogging with sound search for audio retrieval. When the user captures some interesting audio they can immediately upload it to the Web and blog about it, for example, a conversation with a friend. At the end of the day, all the audio is uploaded to the Web for archiving. At a later time, the user may wish to find a particular sound and audio search algorithms are provided to find those interesting moments.

Starting with the basic scenario, a simulation was set up to explore this audio application. One VM represented the wearable device, the second VM a smartphone, and the third VM the server software, as proposed in the last chapter in Figure 2.5. The simulations were enhanced over time to add details and enable moving to real hardware when available. The sensor was modeled initially with a software compression algorithm, then enhanced to use SystemC models with floating-point operations, and, lastly, changed to have SystemC models using fixed-point precision. Open-source SystemC tools from GreenSocs [29] were used to help here. (QEMU-SystemC has been used before to model a MPEG-decoder [61, 62].) The step-wise refinement separated out the critical functions that are the most computationally complex in order to reduce

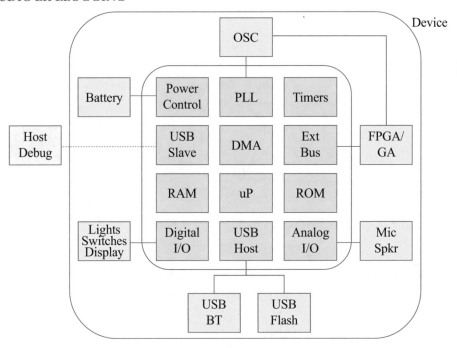

Figure 3.11: Potential block diagram of the sensor electronics.

power. The smartphone code initially developed on the VM Android emulator was ported onto a real smartphone. The server code initially written using a MATLAB-like tool was converted to C++ for more efficient operation and ported to Amazon cloud services. Details follow.

At the outset there were a few basic requirements that had to be met. The sensor must be very low power if it is to be recording for long periods. Therefore, we want to avoid heavy computations and offload them to the server. Continuous recording is going to generate a lot of data and a reliable data link to the server probably can't be guaranteed. So the audio should be stored in the sensor device until needed. Sending the data via a wireless link to the phone or Internet is going to require a relatively power-hungry radio and that task is best done when the sensor is being charged, for example at night while the user is sleeping. We wanted the user to be able to access some audio in real time but limited that to the last recording for immediate blogging purposes. Storing raw audio on the phone consumes a large amount of memory, and so it is best to compress the audio to save space and the time required for the radio to transmit the data. Since compression requires extensive computations we investigated a custom silicon architecture to reduce the power required.

Figure 3.11 shows a potential block diagram of the electronics for the sensor, which is fairly general in nature. The light blue blocks in the center would be part of a standard microcontroller with USB ports and analog I/O. The yellow blocks are other components that connect to the microcontroller. A custom device or FPGA would be required for the audio compression, and other audio operations that come out as requirements from the system analysis. The host debug port to the outside world can be used for software upgrades, testing, and battery charging. A future smartwatch with larger Flash memory for audio storage and DSP for voice compression is an attractive option.

The sensor device may need a local real-time kernel or simple OS to control its operation. For example, to coordinate the operations of the radio, capture of audio data, monitoring of local switches, display updating, and battery monitoring. Figure 3.12 shows a Java-based environment for supporting sensor operation. From the features required, a rough estimate of the size of the control software program and data memory can be determined.

- To simplify the Sensor Device programming, a set of utilities, drivers, and kernel should be provided

- For complex devices, the convenience of Java programming may be worth the memory cost ME/CLDC > 160 KB ROM, > 32 KB RAM

- The user writes the data capture application in C, C++, or Java and downloads it to the platform (real or virtual). For remote control, an external debugger is connected via USB for real HW, or direct for virtual device

Figure 3.12: Sensor software stack.

SIMULATION

As one can imagine, there are many factors to investigate to get to an optimum design. For example, the quality of the compressed audio can greatly impact recognition accuracy. The sound recognition algorithms that work well in a quiet office environment may not work at all in a noisy manufacturing plant. How the recording device is held or attached to the user also impacts the signal quality. The size and weight of the recording device determines the willingness of the user to use the product. What is the best radio to use, should it be a high-speed radio that is only on for a short period, or a slow-speed radio that probably uses less power but takes longer and frustrates the user? How big should the battery be, is operation for one day sufficient, can we employ silence detection to remove unnecessary operations and how much power will that process take? The list goes on and on.

This is where system simulation can help us make some of these trade-offs. The virtual demo gives a better feel for how the system is going to be used in practice. Audio samples can be

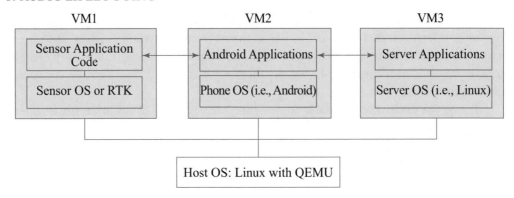

Figure 3.13: Concept-level simulation.

captured offline and fed into the system. The types of usage that people will make of the system will become more apparent, enabling the functionality to be adjusted and tested again.

VMs are used at the initial concept level to test the basic ideas. SystemC and QEMU are used to represent components in the sensor device, The Android emulator is used to model the smartphone, and another instance of QEMU is used to represent the server functionality. QEMU is used under the hood in the Android emulator, thus we have 3 QEMU VMs working together, see Figure 3.13.

From a high-level modeling perspective we can ignore most of the implementation detail! We assume the device captures audio, compresses the data, stores it locally, and transfers the data to the smartphone. The audio compression algorithm is important to model in detail because that must work with the pattern-matching software. A practical implementation of the sensor device must consider the amount of memory required to store the audio recordings, battery life, product cost, and other factors. These factors can be calculated by hand and don't require simulation. It might be worthwhile including a model of the switches and display on the sensor device in the simulations, especially if user interaction is important. Although this was not considered necessary, for a smartwatch implementation it may become more important.

The Android emulator was used to represent the phone software. Since the phone application software is the primary interface into the system and where the user experiences all the features of the system, this level of implementation was deemed important.

The wireless communication links were not modeled, it was assumed they either worked or were disconnected. Since transfer of bulk data occurs when charging and most likely at night, we assume it will eventually get through. A more thorough investigation might implement a wireless link with a probabilistic model to represent disconnect durations.

The server is providing the compute power to implement the audio pattern-matching algorithms and, hence, can be modeled by just the algorithms that run on the server. The server

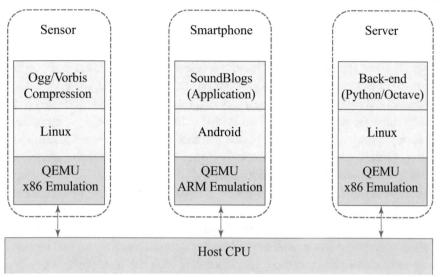

Figure 3.14: Stage 1: Development using three virtual machines.

model is also the interface to the Internet for blogging purposes. This feature was implemented by a connection to the network on the host.

Hence, the high-level model becomes a lot simpler than the future real system, however, it retains enough detail to allow starting the concept exploration. Figure 3.14 shows three virtual machines configured for the audio application. VM1 contains the sensor model, VM2 the phone software, VM3 the server code. Communication between VM's goes via network sockets.

3.4   LIFELOGGING DEVELOPMENT

The project work can be split into three main stages.

1. Stage 1: Concept implementation.

2. Stage 2: Optimizations and Enhancements.

3. Stage 3: Real-world algorithm testing platform.

3.4.1   STAGE 1: CONCEPT IMPLEMENTATION

Stage 1 used three QEMU-based virtual machine instances to represent the sensor, phone, and server. Stage 1 enables exploration of the system software architecture to give a feel for how the system will work in practice.

This arrangement enables blogging of recorded clips, uploading clips to the server for sound classification, and searching the data base for clips of interest. All this is controlled through the

Android application, a first simplified version of the final application. Audio recordings can be loaded into the sensor model, processed on the server, and results shown on the model. The simulation runs slower than real hardware and likely contains bugs. However, that is not a problem at this stage, it is not that hard to imagine the application running faster and bug free.

All three devices are instantiated as VMs on the host machine, see Figure 3.14. The sensor is modeled as an ARM 9 system running a GNU/Linux operating system using QEMU. We have used an open implementation of Ogg/Vorbis [63] as the compression algorithm. The source code for the encoder is run on the virtual machine instance without any modifications and additional coprocessors. Once compressed, the audio is stored in memory and the sensor awaits further instructions from the smartphone for data transfer.

An Android emulator is used to model a smartphone capable of emulating GPS-based locations and wireless data transfer.

For the server model, we used QEMU to emulate a full-fledged GNU/Linux server responsible for serving up web pages as well as performing a set of classification and retrieval algorithms on the user-uploaded audio recordings. As is the case with the development of algorithms, they were first written in a high-level language such as MATLAB, for simulation and testing purposes. To illustrate the ability of QEMU in handling this aspect, we used an open-source counterpart of MATLAB, called GNU/Octave [64], to implement the algorithms on the server model.

During development of the Android application in the concept implementation stage it was realized that users would benefit from a map display of where recordings were made. When the recordings are made, the GPS geographic coordinates of the phone are stored as part of the audio data. The software was enhanced to show the place of recording and identified on the map with a colored icon. A related enhancement enables the user to see what other recording icons are present in the vicinity of the first icon. So, for example, if a audio recording made while on vacation was found and shown on the map near the beach, the user could browse around the map and listen to other recordings made during the same vacation.

Stage 1 Operation

The virtual system comprising the three devices instantiated separately on a single host machine is shown in Figure 3.15. The sequence of actions to perform blogging using this framework can be explained as follows. An audio clip is loaded into the virtual sensor from a separately captured audio recording. The audio encoder program is called to compress the clip and store it in a file. This clip is then simultaneously uploaded to the virtual smartphone and the virtual server. The Android application on the smartphone displays this clip, allowing a user to listen and add notes. Simultaneously, the server processes the data and automatically annotates it with relevant tags. Once the user decides to archive/publish this event, all the details, including the audio clip, tags, description, are packaged in a single file and uploaded to the user's blog website or personal archive. A similar process exists for the retrieval of clips from a user's personal database.

Figure 3.16 shows event mapping included in the Android application.

Figure 3.15: Stage 1: Linux windows for each virtual machine.

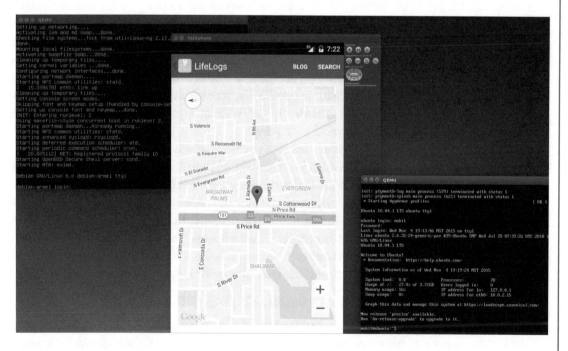

Figure 3.16: Screenshot showing mapping feature.

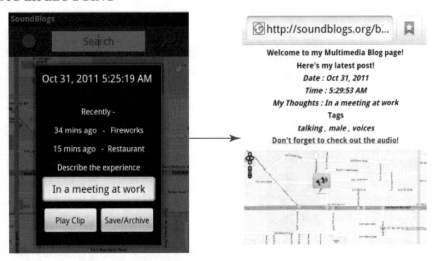

Figure 3.17: Blogging operation.

Clicking the blog button in the Android application triggers a script to transfer a file with uncompressed audio to the sensor virtual machine. The encoder running on the emulated ARM processor generates an audio file. It uses the Linux ssh command to transfer the compressed audio file to the virtual machine running the Android emulator. The data is then forwarded to the server for searching the audio database. To do this, a Python script runs on the host looking for file updates and forwards them to server or sensor as selected by filename. After the server has identified the sound tags associated with the audio file, it sends the tags back to the phone using the Linux ssh command for display to the user in the Android application. For example, recorded sounds of people talking will be decoded as "talking."

Clicking the search button prompts the user to enter a search term and runs another transfer script. The entry is written to a file, which is seen by the Python script and sent to the server. The server matches the search term with labeled recordings associated with the user in the database and returns them to the phone. On the phone, the user can see on the map where those retrieved recordings were made. For example, the search term "alarm" might find a recording of a fire alarm going off and will return that file and its geographic coordinates. The phone then displays location icons on the map and clicking on an icon will replay the audio file.

Figures 3.17 and 3.18 show screenshots from the application in action.

3.4.2   STAGE 2: OPTIMIZATIONS AND ENHANCEMENTS

Stage 2 starts to explore the hardware architecture in the sensor module. The transform operations involve heavy number crunching and are pulled out as a separate block. Stage 2 can be the starting point for developing the software needed to control the sensor device, although that is not explored here. The server code is converted to C++ to improve performance.

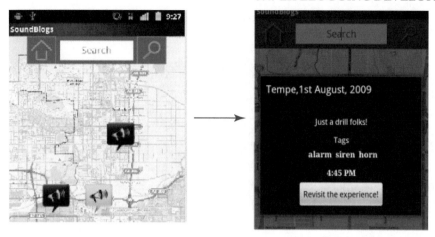

Figure 3.18: Search operation.

For the sensor, we are interested in building a wearable device that is non-obstructive, small, and light weight. The device should be able to record, compress, and store audio for up to 24 hours without having to recharge the batteries. In order to satisfy this requirement, we must ensure that the device operates at ultra-low power. Hence, changes are made to the compression algorithm to reduce the computational complexity as much as possible.

Before we perform any kind of optimizations on the compression algorithm, we must first identify its computationally intensive routines. A time profile of the Ogg/Vorbis encoder and the sub-routines is shown in Figure 3.19. These routines include the Fast Fourier Transform (FFT) [65], Modified Discrete Cosine Transform (MDCT) [52], Tone-Masking Threshold (TMT), Noise-Masking Threshold (NMT), and miscellaneous (MISC) operations related to Huffman encoding and packing the encoded data. Although TMT and NMT constitute 54.1% of the total time, they involve a large number of comparison and threshold operations that are inherently simple to implement and cannot be optimized further. The FFT and MDCT routines, taking 30% of the total time, rely on a significant amount of real and complex multiplication operations.

We developed virtual hardware devices or coprocessors using SystemC to perform MDCT, FFT, TMT, and NMT. In this stage, all functions were implemented using floating-point arithmetic. At first, these devices were tested in a pure SystemC simulation environment. A traffic generator sends bursts of data to a generic router. The data is routed to the selected coprocessor for further computations. A virtual device emulating a ROM is also designed to store sets of coefficients used in FFT and MDCT operations. These blocks are then plugged into QEMU-SystemC through the SystemC interface. The Ogg/Vorbis application code was modified to transfer these operations to the peripheral blocks.

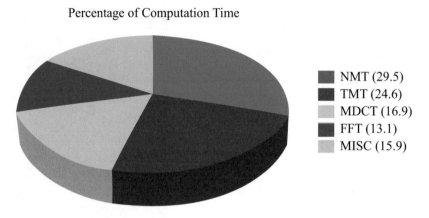

Figure 3.19: Functional time profile.

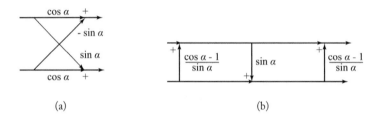

Figure 3.20: Lifting operation.

Our next step was aimed at modifying the floating-point FFT and MDCT routines to reduce the overall power consumption. The multiplication operations involved in computing the MDCT and FFT were based on the popular butterfly unit. In [66], Oraintara et al. derived a lifting-based method to convert these units to lattice structure's as shown in Figure 3.20. This method allows for quantization and perfect reconstruction without considerable loss of precision. The trigonometric coefficients employed in these computations can be stored as dyadic rational numbers [67]. As a result, multiplications can now be implemented as a series of shift-add operations. We specify the input data and coefficients as 16-bit and 10-bit signed integers respectively. The reconstruction error based on these specifications, as given in [66], is approximately -100 dB. The error is low enough to preserve the quality of recorded audio. Once again, we tested these devices in a pure SystemC simulation environment, followed by QEMU-SystemC.

In the first stage of the server model, the algorithms for pattern recognition, annotation, and retrieval are implemented in a highly functional language, i.e., MATLAB or GNU/Octave. This language suffices for initial testing, however, there is a significant overhead due to its higher abstraction, making it very slow. Hence, the algorithms were ported to C/C++ for better speed and memory performance.

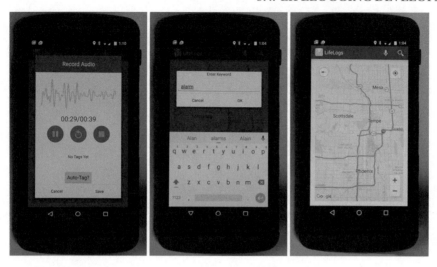

Figure 3.21: Lifelogging application running on an Android Phone.

3.4.3   STAGE 3: REAL-TIME ALGORITHM TESTING

Stage 3 uses the phone and cloud processing as a separate platform to capture more environmental sounds. This is for refining the sensor preprocessing algorithms, sound recognition algorithms, and the user interface in the real world.

The phone application was ported from the QEMU-based Android emulator to an actual smartphone to enable real-time location capture. The server algorithms were ported to the Amazon EC2 cloud-based server. This allows for users with the Android application to archive and access their personal recordings from anywhere in the world.

In this last stage, we bypassed a separate sensor and chose to record the audio on the phone. Switching to real hardware at this point allowed more refinement of the phone-user interface and algorithm optimization on the server, which would be difficult in the virtual domain. More data collection in real-world conditions is needed to tune the algorithms to be more robust in noisy and varied environments. With the application being portable on a phone, the data collection and algorithm work can proceed quickly. The phone can be taken out into the streets, countryside, or wherever, to record environmental sounds and immediately test the search algorithms. It is not necessary to have the sensor at this stage for that work to proceed.

The smartphone application code is modified to call the Android audio capture library routines. The server code stays largely unchanged. Code has been added to help with statistical analysis of the data collection process.

Figure 3.21 shows a more recent version of the application running on an Android phone.

CHAPTER 4

Summary

4.1 A DISCUSSION OF THE AUDIO LIFELOGGING PROJECT

The audio project demonstrated that a novel idea could be explored in the virtual world and many details implemented in advance of hardware availability. The Android application shows what can be expected in a real-world implementation. The software developed for the sensor and server are a good starting point for continuing that part of the project through to a product.

The map feature showing where recordings were located came out of the concept investigation and greatly influenced the look and feel of the Android application. This provided a nice addition to the user interface and made the concept more attractive.

There is much more that can be done in the virtual world to continue the project through to a real product. After field trials with the Stage 3 phone-based implementation, further enhancements may be necessary, such as noise cancellation. Once the requirements are complete, a sensor DSP co-processor architecture and microcode can then be designed and simulated to optimize power and performance.

Rather than Linux, a more compact real-time kernel for management of various control and I/O functions in the sensor could reduce cost. The functional virtual platform is well suited for porting and debugging that code.

The project only considered a single user, support for multiple users on the server is needed for full deployment. The software should be scalable so that more VM's in the cloud can be added as demand ramps up. A scalable software architecture using containers could be developed to handle the growth in users.

However, at some point it will make sense to jump to hardware, since it will be faster than simulation. Also, there are going to be real-world problems, which either cannot be simulated or were not thought about earlier. While the hardware is being built, more application software details can be added and run on the simulators.

Further research to improve accuracy could be facilitated using the virtual implementation. For example, geographic coordinates that are recorded to show where the audio recording is being made could be used to identify the surrounding environment, i.e., city center, at home, or at a peaceful lake. Sensor fusing techniques [68, 69] can be used to include knowledge of the environment to improve the sound recognition accuracy.

So in summary, a novel audio recording device was explored in the virtual world, problems discovered, application features improved, and ideas for future enhancements uncovered. Much can be done before needing hardware.

4.2   DISCUSSION OF METHODOLOGY

This book has focused on IoT applications that primarily involve personal devices, in our case audio sensors and phones connected to cloud computing. Other larger projects may need more powerful software tools, such as those available from suppliers discussed in Section 2.11. However, the underlying theme is to do as much as possible top down to find system problems first and then incrementally add detail to get to real hardware.

The transition to a top-down methodology from a bottom-up approach has to take into account a desire to build hardware as soon as possible. Preparing models and software does takes time that could have been spent elsewhere. One has to remember that systems issues being resolved in modeling and simulation could have been catastrophic at a later stage if left undiscovered. Plus, much of the application software developed on the models ends up in the hardware.

It should be noted that it is not necessary to simulate at the concept level everything the device might have to deal with. Many of the problems and potential enhancements will become clear from just implementing the basic features. It is the demonstration of the idea that gets one to consider all the implications and possible enhancements. Missing details can be added as the concept model is transformed into the virtual platform.

The virtual platform is not redundant once the hardware is built and the services deployed. Some complex bugs are easier to diagnose using simulation, and there are always future improvements to investigate. For example, video logging could be added to the concept models and virtual platform.

Ideally, most of the design and development of hardware and software systems would be done in the virtual world. Problems found in the real world after hardware is built are noted and used to refine the simulations, so that next time around the design loop even more can be done in the virtual world.

4.3   CONCLUSIONS

The audio project has demonstrated that it is possible to design hardware and software in the virtual space by making use of modern simulation techniques. Systems and chips are so complex these days that it is important to find problems early before committing to expensive hardware. Simulation starting at the concept level naturally leads to a top-down approach where more details get added in stages. By starting at the highest level, important system issues have less chance of being overlooked and are easier to fix. We have made use of virtual machine and virtual platform technology to enable this approach and much of a design can be conveniently developed on a PC using these tools. Open-source software was used to develop a fairly sophisticated audio application showing that much can be done at low cost.

The material here just scratches the surface of what is possible, much more can done with simulation and design automation. Several companies are developing software tools to make this process easier and the next few years should see some exciting new tools. However, if you want to start now at little or no expense, then check out the open-source software described in this book. That too will improve over time.

Bibliography

[1] L. Atzori, A. Iera, and G. Morabito, "The internet of things: A survey," *Computer Networks*, vol. 54, no. 15, pp. 2787–2805, 2010. DOI: 10.1016/j.comnet.2010.05.010. 1

[2] J. Gubbia, R. Buyyab, S. Marusica, and M. Palaniswamia, "Internet of things (IoT): A vision, architectural elements, and future directions," *Future Generation Computer Systems*, vol. 29, no. 7, pp. 1645–1660, 2013.

[3] L. D. Xu, W. He, and S. Li, "Internet of things in industries: A survey," *IEEE Transactions on Industrial Informatics*, vol. 10, pp. 2233–2243, Nov 2014. DOI: 10.1109/tii.2014.2300753. 1

[4] A. Zaslavsky, "Internet of things and ubiquitous sensing," *Computing Now*, September 2013. 3

[5] R. P. Goldberg, "A survey of virtual machine research," *Computer*, vol. 7, no. 6, pp. 34–45, 1974. DOI: 10.1109/mc.1974.6323581. 5

[6] Y. Li, W. Li, and C. Jiang, "A survey of virtual machine system: Current technology and future trends," in *Electronic Commerce and Security (ISECS), 2010 Third International Symposium on*, pp. 332–336, July 2010. DOI: 10.1109/isecs.2010.80. 5

[7] "Cadence Incisive Enterprise Simulator." http://www.cadence.com/products/fv/en terprise_simulator/pages/default.aspx 5, 27

[8] "Carbon Design Systems SoC Designer Plus." http://www.carbondesignsystems.co m/soc-designer-plus 5, 27

[9] A. Barreteau, "System-level modeling and simulation with Intel® CoFluentTM studio," in *Complex Systems Design and Management, Proceedings of the Sixth International Conference on Complex Systems Design and Management*, pp. 305–306, Springer, 2015. DOI: 10.1007/978-3-319-26109-6_32. 5, 15

[10] D. Aarno and J. Engblom, *Software and System development Using Virtual Platforms*. Morgan Kaufmann, 2015. 5

[11] MentorGraphics, "MentorGraphics Vista Virtual Prototyping." https://www.mentor.c om/esl/vista/virtual-prototyping/ 5, 27

[12] "Synopsys Virtual Prototyping." `http://www.synopsys.com/Prototyping/VirtualPr` `ototyping/Pages/default.aspx` 5, 27

[13] I. S. 1666-2011, "IEEE Standard for Standard SystemC Language Reference Manual," tech. rep., 2011. Available from `https://standards.ieee.org/findstds/standard` `/1666-2011.html`. DOI: 10.1109/ieeestd.2012.6134619. 6, 23

[14] B. Bentley, "Validating the Intel® Pentium® 4 Microprocessor," *DAC 2001, June 18–22, Las Vegas, NV*, 2001. 9

[15] G. Martin and H. Chang, *Winning the SoC Revolution, Experiences in Real Design*. Springer Science+Business Media LLC, 2003. 9

[16] M. Keating, D. Flynn, R. Aitken, A. Gibbons, and K. Shi, *Low Power Methodology Manual: For System-on-Chip Design*. Springer Publishing Company, Incorporated, 2007. DOI: 10.1007/978-1-4757-2887-3. 10

[17] A. P. Chandrakasan and R. W. Brodersen, *Low Power Digital CMOS Design*. Springer Science+Business Media LLC, 1995. DOI: 10.1007/978-1-4615-2325-3. 10

[18] S. Kinney, *Trusted Platform Module Basics: Using TPM in Embedded Systems*. Elsevier Inc, 2006. 10

[19] R. Elbaz, L. Torres, G. Sassatelli, P. Guillemin, C. Anguille, C. Buatois, and J. B. Rigaud, "Hardware engines for bus encryption: a survey of existing techniques," in *Design, Automation and Test in Europe, 2005. Proceedings of*, pp. 40–45 Vol. 3, March 2005. DOI: 10.1109/date.2005.170. 10

[20] G. Asada, M. Dong, T. S. Lin, F. Newberg, G. Pottie, W. J. Kaiser, and H. O. Marcy, "Wireless integrated network sensors: Low power systems on a chip," in *Solid-State Circuits Conference, 1998. ESSCIRC '98. Proc. of the 24th European*, pp. 9–16, Sept 1998. DOI: 10.1109/ESSCIR.1998.186200. 10

[21] A. A. Abidi, "Direct-conversion radio transceivers for digital communications," *IEEE Journal of Solid-State Circuits*, vol. 30, pp. 1399–1410, Dec 1995. DOI: 10.1109/4.482187. 10

[22] M. H. Zaki, S. Tahar, and G. Bois, "Formal verification of analog and mixed signal designs: A survey," *Microelectronics Journal*, vol. 39, no. 12, pp. 1395–1404, 2008. DOI: 10.1016/j.mejo.2008.05.013. 10

[23] M. Fowler, *UML Distilled* 3rd ed., Addison Wesley, 2004. 15

[24] QEMU. `www.qemu.org` 16

[25] F. Bellard, "QEMU, a fast and portable dynamic translator," *USENIX Annual Technical Conference, FREENIX Track*, pp.41–46, 2005. 16

[26] "KVM." www.linux-kvm.org 17

[27] A. V. Romero, *VirtualBox 3.1: Beginner's Guide.* Packt Publishing, 2010. 17

[28] R. Troy and M. Helmke, *VMWare Cookbook*, 2nd ed., O'Reilly, 2012. 17

[29] "GreenSocs." www.greensocs.com 17, 27, 41

[30] "GreenSocs QBOX." https://git.greensocs.com/qemu/qbox 17, 27

[31] "Linux Containers." https://linuxcontainers.org 17

[32] F. Ghenassia, *Transaction-level Modeling with SystemC.* Springer Publishing, 2005. DOI: 10.1007/b137175. 23

[33] D.C. Black, J. Donovan, B. Bunton, and A. Keist, *SystemC: From the Ground Up.* Springer Publishing, 2009. DOI: 10.1007/978-0-387-69958-5.

[34] T. Grotker, S. Liao, G. Martin, and S. Swan, *System Design with SystemC.* Springer Publishing, 2007. DOI: 10.1007/b116588.

[35] B. Bailey and G. Martin, *ESL Models and their Application: Electronic System Level Design and Verification in Practice.* Springer Publishing, 2009. 23

[36] M. Burton and A. Morawiec, *Platform Based Design at the Electronic System Level: Industry Perspectives and Experiences.* Springer Publishing, 2007. DOI: 10.1007/1-4020-5138-7. 27

[37] B. P. Rimal, E. Choi, and I. Lumb, "A taxonomy and survey of cloud computing systems," *Networked Computing and Advanced Information Management, International Conference on,* vol. 0, pp. 44–51, 2009. DOI: 10.1109/ncm.2009.218. 28

[38] J. van Vliet and F. Paganelli, *Programming Amazon EC2.* Springer Publishing, 2011. 28

[39] "Microsoft Azure IoT Suite." https://www.microsoft.com/en-us/server-cloud/internet-of-things/azure-iot-suite.aspx 29

[40] "IBM Bluemix." https://console.ng.bluemix.net/ 29

[41] V. Bush, "As we may think." *The Atlantic Monthly*, 1945. DOI: 10.1145/227181.227186. 31

[42] D. P. W. Ellis and K. S. Lee, "Minimal-impact audio-based personal archives," *1st ACM Workshop Continuous Archival and Retrieval of Personal Experiences*, ACM Press, 2004. DOI: 10.1145/1026653.1026659.

[43] A. R. Doherty, N. Caprani, C. A. Conaire, V. Kalnikaite, C. Gurrin, A. F. Smeaton, and N. E. O'Connor, "Passively recognising human activities through lifelogging," *Computers in Human Behavior*, vol. 27, no. 5, pp. 1948–1958, 2011. 2009 Fifth International Conference on Intelligent Computing ICIC 2009. DOI: 10.1016/j.chb.2011.05.002.

[44] B. Clarkson, K. Mase, and A. Pentland, "The familiar: A living diary and companion," *Proc. ACM Conf. Computer–Human Interaction*, pp. 271–272, 2006. DOI: 10.1145/634225.634228.

[45] M. Blum, A. Pentland, and G. Tröster, "InSense: Interest-based Life logging," *IEEE Computer Society*, 2016. DOI: 10.1109/mmul.2006.87.

[46] J. Gemmell, G. Bell, R. O'Reilly MediaLueder, S. Drucker, and C. Wong, "MyLifeBits: fulfilling the Memex vision," in *Proc. of the 10th International Conference on Multimedia*, pp.235–238, 2002. DOI: 10.1145/641007.641053. 31

[47] B. Clarkson, N. Sawhney, and A. Pentland, "Auditory context awareness via wearable computing, in " *Perceptual User Interfaces Workshop*, ACM Press, 1998. 31

[48] R. Weber, "Internet of things—New security and privacy challenges," *Computer Law and Security Review*, Vol. 26, Issue 1, pp. 23–30, 2010. DOI: 10.1016/j.clsr.2009.11.008. 32

[49] M. Shah, B. Mears, C. Chakrabarti, and A. Spanias, "Lifelogging: Archival and retrieval of continuously recorded audio using wearable devices," *Emerging Signal Processing Applications (ESPA) 2012, IEEE International Conference on*, 2012. DOI: 10.1109/espa.2012.6152455. 33

[50] M. Shah, B. Mears, C. Chakrabarti, and A. Spanias, "A top-down design methodology using virtual platforms for concept development," *Quality Electronic Design (ISQED) 2012, 13th International Symposium on*, 2012. DOI: 10.1109/isqed.2012.6187531. 33

[51] A. S. Spanias, "Speech coding: A tutorial review," *Proc. of the IEEE*, vol. 82, no. 10, pp. 1541–1582, 1994. DOI: 10.1109/5.326413. 33

[52] A. S. Spanias, T. Painter, and V. Atti, *Audio Signal Processing and Coding*. Wiley, March 2007, hardcover. DOI: 10.1002/0470041978. 49

[53] T. Painter and A. Spanias, "Perceptual coding of digital audio," *Proc. of the IEEE*, vol. 88, no. 4, pp. 451–515, 2000. DOI: 10.1109/5.842996.

[54] K. Brandenburg and G. Stoll, "ISO/MPEG-1 audio: A generic standard for coding of high-quality digital audio," *J. Audio Eng. Soc.*, vol. 42, no. 10, pp. 780–792, 1994.

[55] M. Bosi, K. Brandenburg, S. Quackenbush, L. Fielder, K. Akagiri, H. Fuchs, and M. Dietz, "ISO/IEC MPEG-2 advanced audio coding," *J. Audio Eng. Soc.*, vol. 45, no. 10, pp. 789–814, 1997.

[56] A. Spanias, M. Deisher, P. Loizou, G. Lim, and B. Mears, "A new highly integrated architecture for speech processing and communication applications," *Intel Technology Journal*, vol. Spring, pp. 41–56, 1994. 33

[57] G. Wichern, H. Thornburg, B. Mechtley, A. Fink, K. Tu, and A. Spanias, "Robust multi-features segmentation and indexing for natural sound environments," *IEEE International Workshop on Content-Based Multimedia Indexing, pp. 69–76*, 2007. DOI: 10.1109/cbmi.2007.385394. 34

[58] G. Wichern, J. Xue, H. Thornburg, B. Mechtley, and A. Spanias, "Segmentation, indexing, and retrieval for environmental and natural sounds," *EEE Transactions on Audio, Speech, and Language Processing*, vol. 18, no. 3, pp. 688–707, 2010. DOI: 10.1109/tasl.2010.2041384. 34

[59] K. P. Murphy, *Dynamic Bayesian Networks: Representation, Inference and Learning*. Ph.D. thesis, University of California, Berkeley, 2002. 37

[60] T. Pedersen, S. Patwardhan, and J. Michelizzi, "WordNet::Similarity: measuring the relatedness of concepts," *Proceeding HLT-NAACL–Demonstrations '04 Demonstration Papers at HLT-NAACL 2004*, pp. 38–41, 2004. DOI: 10.3115/1614025.1614037. 39

[61] K. Gruttner, F. Oppenheimer, W. Nebel, F. Colas-Bigey, and A. Fouilliart, "SystemC-based modeling, seamless refinement, and synthesis of a JPEG 2000 decoder," *Design Automation and Test in Europe (DATE)*, pp. 128–133, 2008. DOI: 10.1109/date.2008.4484674. 41

[62] J. J. Thiagarajan and A. Spanias, *Analysis of the MPEG-1 Layer III (MP3) Algorithm Using MATLAB*. Morgan and Claypool Publishers, Synthesis Lectures on Algorithms and Software in Engineering, November 2011. Vol. 3, No. 3, pp. 1–129. DOI: 10.2200/s00382ed1v01y201110ase009. 41

[63] Ogg/Vorbis. http://www.xiph.org/vorbis 46

[64] J. S. Hansen, *GNU Octave Beginner's Guide*. Packt Publishing, 2011. 46

[65] A. Spanias, *Digital Signal Processing; An Interactive Approach*, 2nd ed., Lulu Press On-demand Publishers Morrisville, NC, May 2014. Textbook with JAVA exercises. 49

[66] S. Oraintara, Y. Chen, and T. Nguyen, "Integer Fast Fourier transform," *IEEE Transactions on Signal Processing*, vol. 50, no. 3, pp. 607–618, 2002. DOI: 10.1109/78.984749. 50

[67] T. Tran, "The BinDCT: fast multiplierless approximation of the DCT," *IEEE Signal Processing Letters*, vol. 7, pp. 141–145, 2000. DOI: 10.1109/97.844633. 50

[68] C. Thomas, *Sensor Fusion and its Applications*. Sciyo, 2010. DOI: 10.5772/3302. 53

[69] M. Stanley, "Sensor fusion, nxp." http://blog.nxp.com/author/michael-stanley/ 53

Authors' Biographies

BRIAN MEARS

Brian Mears has a Ph.D. from City University, London. He worked in the computer industry in the U.K., at CERN in Geneva, at Bell Labs designing integrated circuits for audio compression, and at Intel Corporation for 29 years where he was a silicon architect and manager for embedded 32-bit microcontrollers, multi-core digital signal processors, and smartphone SoCs. His experience with complex SoC chip design led him to employ modeling and simulation techniques for silicon architecture design and software development. Retired from Intel Corporation he is currently an adjunct faculty member at Arizona State University researching design techniques and tools for architecture design.

MOHIT SHAH

Mohit Shah received his Ph.D. from Arizona State University in 2015. He co-founded Genesis Artificial Intelligence, which specializes in advanced system and application design. He received his B.Tech from Nirma University, India in 2008. His research interests lie in speech processing and analysis methods, covering both DSP and machine-learning related aspects. He has authored several publications in these areas. His prior work in industry included research internships at Intel and Palo Alto Research Center (Xerox PARC) in 2012 and 2013, respectively.